T0210483

Lecture Notes in Computer Science 9061

Commenced Publication in 1973
Founding and Former Series Editors:
Gerhard Goos, Juris Hartmanis, and Jan van Leeuwen

More information about this series at http://www.springer.com/series/7407

Çetin Kaya Koç · Sihem Mesnager
Erkay Savaş (Eds.)

Arithmetic
of Finite Fields

5th International Workshop, WAIFI 2014
Gebze, Turkey, September 27–28, 2014
Revised Selected Papers

 Springer

Editors
Çetin Kaya Koç
Department of Computer Science
University of California, Santa Barbara
Santa Barbara, CA
USA

Erkay Savaş
Sabancı University
Istanbul
Turkey

Sihem Mesnager
University of Paris VIII
Paris
France

ISSN 0302-9743   ISSN 1611-3349 (electronic)
Lecture Notes in Computer Science
ISBN 978-3-319-16276-8   ISBN 978-3-319-16277-5 (eBook)
DOI 10.1007/978-3-319-16277-5

Library of Congress Control Number: 2015932669

LNCS Sublibrary: SL1 – Theoretical Computer Science and General Issues

Springer Cham Heidelberg New York Dordrecht London

Printed on acid-free paper

Springer International Publishing AG Switzerland is part of Springer Science+Business Media
(www.springer.com)

Preface

This volume contains the revised and expanded versions of the papers presented at the 5th International Workshop on the Arithmetic of Finite Fields (WAIFI). The workshop was held in Gebze, Turkey, during September 27–28, 2014.

The final program contained three invited and nine contributed papers, which are all found in this volume. The contributed papers were selected from 17 submissions using a careful refereeing process. At least three and in most cases four referees reviewed each paper. The final decisions were taken only after a clear position could be clarified through additional reviews and comments.

We are very grateful and express our thanks to the Program Committee Chairs, the Program Committee Members, and also to the external reviewers for their great work of reviewing. Their help and cooperation was essential, especially due to the short amount of time available to conduct the reviewing task.

The Program Committee invited Claude Carlet, Ferruh Özbudak, and Francisco Rodríguez-Henríquez to speak on topics of their choice, as related to the WAIFI 2014. We thank them for having accepted. Their contributions provided a valuable framing for the contributed papers.

The Steering Committee also thanks the General Chair, Çetin Kaya Koç, and the Program Co-chairs, Sihem Mesnager and Erkay Savaş for the rigorous work. Furthermore, the Committee thanks Jean-Jacques Quisquater and José Luis Imaña for their valuable help in publicity and web page matters.

Springer has published all previous volumes of the WAIFI Proceedings:

- Ferruh Özbudak and Francisco Rodríguez-Henríquez (Eds.): Arithmetic of Finite Fields, 4th International Workshop, WAIFI 2012, Bochum, Germany, July 16–19, 2012. Springer, LNCS Volume 7369.
- M. Anwar Hasan and Tor Helleseth (Eds.): Arithmetic of Finite Fields, 3rd International Workshop, WAIFI 2010, Istanbul, Turkey, June 27–30, 2010. Springer, LNCS Volume 6087.
- Joachim von zur Gathen, José Luis Imaña, and Çetin Kaya Koç (Eds.): Arithmetic of Finite Fields, 2nd International Workshop, WAIFI 2008, Siena, Italy, July 6–9, 2008. Springer, LNCS Volume 5130.
- Claude Carlet and Berk Sunar (Eds.): Arithmetic of Finite Fields, 1st International Workshop, WAIFI 2007, Madrid, Spain, June 21–22, 2007. Springer, LNCS Volume 4547.

As with the previous volumes, Springer agreed to publish the revised and expanded versions of the WAIFI 2014 papers as an LNCS volume. We thank Alfred Hoffman and Ronan Nugent from Springer for making this possible.

September 2014

Çetin Kaya Koç
Sihem Mesnager
Erkay Savaş

Organization

Committees

Steering Committee

Claude Carlet	University of Paris VIII, France
Jean-Pierre Deschamps	Rovira i Virgili University, Spain
José Luis Imaña	Complutense University of Madrid, Spain
Çetin Kaya Koç	University of California, Santa Barbara, USA
Ferruh Özbudak	Middle East Technical University, Turkey
Christof Paar	Ruhr University Bochum, Germany
Jean-Jacques Quisquater	Université catholique de Louvain, Belgium
Francisco Rodríguez-Henríquez	CINVESTAV-IPN, Mexico
Berk Sunar	Worcester Polytechnic Institute, USA
Gustavo Sutter	Autonomous University of Madrid, Spain

General Chair

Çetin Kaya Koç	University of California, Santa Barbara, USA

Program Chairs

Sihem Mesnager	University of Paris VIII, France
Erkay Savaş	Sabancı University, Turkey

Publicity Chair

Jean-Jacques Quisquater	Université catholique de Louvain, Belgium

Local Arrangements Chairs

Şükran Külekçi	Tübitak BİLGEM, Turkey
Mehmet Sabır Kiraz	Tübitak BİLGEM, Turkey

Program Committee

Daniel Augot	Inria and LIX, France
Lejla Batina	Radboud University Nijmegen, The Netherlands
Luca Breveglieri	Politecnico di Milano, Italy
Claude Carlet	University of Paris VIII, France
Murat Cenk	Middle East Technical University, Turkey
Gérard Cohen	Télécom ParisTech, France
Philippe Gaborit	University of Limoges, France
Pierrick Gaudry	CNRS, Nancy, France
Tor Helleseth	University of Bergen, Norway

Hüseyin Hışıl   Yaşar University, Turkey
Mehran Mozaffari Kermani   Rochester Institute of Technology, USA
Alexander Kholosha   University of Bergen, Norway
Gregor Leander   Ruhr University Bochum, Germany
Julio López   University of Campinas, Brazil
Wilfried Meidl   Sabancı University, Turkey
Sihem Mesnager   University of Paris VIII, France
Christophe Negre   Université de Perpignan, France
Harald Niederreiter   RICAM, Austrian Academy of Sciences, Austria
Erdinç Öztürk   Istanbul Commerce University, Turkey
Alexander Pott   Otto-von-Guericke University, Germany
Arash Reyhani-Masoleh   University of Western Ontario, Canada
Francisco Rodríguez-Henríquez   CINVESTAV-IPN, Mexico
Erkay Savaş   Sabancı University, Turkey
Zülfükar Saygı   TOBB Ekonomi ve Teknoloji Üniversitesi, Turkey
Kai-Uwe Schmidt   Otto-von-Guericke University, Germany
Leo Storme   Ghent University, Belgium
Jean-Pierre Tillich   Inria-Rocquencourt, France

Additional Reviewers

Çetin Kaya Koç   University of California, Santa Barbara, USA
Jean-Jacques Quisquater   Université catholique de Louvain, Belgium

Contents

Third Invited Talk

Coding Theory and Code-Based Cryptography

First Invited Talk

Computing Discrete Logarithms in $\mathbb{F}_{3^{6 \cdot 137}}$ and $\mathbb{F}_{3^{6 \cdot 163}}$ Using Magma

Gora Adj[1], Alfred Menezes[2], Thomaz Oliveira[1],
and Francisco Rodríguez-Henríquez[1(✉)]

[1] Computer Science Department, CINVESTAV-IPN, Mexico City, Mexico
{gora.adj,thomaz.figueiredo}@gmail.com,
francisco@cs.cinvestav.mx
[2] Department of Combinatorics and Optimization, University of Waterloo,
Waterloo, Canada
ajmeneze@uwaterloo.ca

Abstract. We show that a Magma implementation of Joux's $L[1/4 + o(1)]$ algorithm can be used to compute discrete logarithms in the 1303-bit finite field $\mathbb{F}_{3^{6 \cdot 137}}$ and the 1551-bit finite field $\mathbb{F}_{3^{6 \cdot 163}}$ with very modest computational resources. Our $\mathbb{F}_{3^{6 \cdot 137}}$ implementation was the first to illustrate the effectiveness of Joux's algorithm for computing discrete logarithms in small-characteristic finite fields that are not Kummer or twisted-Kummer extensions.

1 Introduction

Let $\mathbb{F}_Q$ denote the finite field of order Q. The discrete logarithm problem (DLP) in $\mathbb{F}_Q$ is that of determining, given a generator g of $\mathbb{F}_Q^*$ and an element $h \in \mathbb{F}_Q^*$, the integer $x \in [0, Q - 2]$ satisfying $h = g^x$. In the remainder of the paper, we shall assume that the characteristic of $\mathbb{F}_Q$ is 2 or 3.

Until recently, the fastest general-purpose algorithm known for solving the DLP in $\mathbb{F}_Q$ was Coppersmith's 1984 index-calculus algorithm [9] with a running time of $L_Q[\frac{1}{3}, (32/9)^{1/3}] \approx L_Q[\frac{1}{3}, 1.526]$, where as usual $L_Q[\alpha, c]$ with $0 < \alpha < 1$ and $c > 0$ denotes the expression $\exp\left((c + o(1))(\log Q)^\alpha (\log \log Q)^{1-\alpha}\right)$ that is subexponential in $\log Q$. In February 2013, Joux [23] presented a new DLP algorithm with a running time of $L_Q[\frac{1}{4} + o(1), c]$ (for some undetermined c) when $Q = q^{2n}$ and $q \approx n$. Shortly thereafter, Barbulescu, Gaudry, Joux and Thomé [4] presented an algorithm with *quasi-polynomial* running time $(\log Q)^{O(\log \log Q)}$ when $Q = q^{2n}$ with $q \approx n$.

These dramatic developments were accompanied by some striking computational results. For example, Gölöğlu et al. [16] computed logarithms in $\mathbb{F}_{2^{8 \cdot 3 \cdot 255}} = \mathbb{F}_{2^{6120}}$ in only 750 CPU hours, and Joux [24] computed logarithms in $\mathbb{F}_{2^{8 \cdot 3 \cdot 257}} = \mathbb{F}_{2^{6168}}$ in only 550 CPU hours. The small computational effort expended in these experiments depends crucially on the special nature of the fields $\mathbb{F}_{2^{6120}}$ and $\mathbb{F}_{2^{6168}}$ — namely that $\mathbb{F}_{2^{6120}}$ is a degree-255 extension of $\mathbb{F}_{2^{8 \cdot 3}}$ with $255 = 2^8 - 1$ (a Kummer extension), and $\mathbb{F}_{2^{6168}}$ is a degree-257 extension of $\mathbb{F}_{2^{8 \cdot 3}}$ with

© Springer International Publishing Switzerland 2015
Ç. Koç et al. (Eds.): WAIFI 2014, LNCS 9061, pp. 3–22, 2015.
DOI: 10.1007/978-3-319-16277-5_1

$257 = 2^8 + 1$ (a twisted Kummer extension). Adj et al. [1] presented a concrete analysis of the new algorithms and demonstrated that logarithms in $\mathbb{F}_{3^{6 \cdot 509}}$ can be computed in approximately 2^{82} time, which is considerably less than the 2^{128} time required by Coppersmith's algorithm. Adj et al. [2] also showed how a modification of the new algorithms by Granger and Zumbrägel [21] can be used to compute logarithms in $\mathbb{F}_{3^{6 \cdot 1429}}$ in approximately 2^{96} time, which is considerably less than the 2^{192} time required by Coppersmith's algorithm. Unlike the aforementioned experimental results, the analysis by Adj et al. does not exploit any special properties of the fields $\mathbb{F}_{3^{6 \cdot 509}}$ and $\mathbb{F}_{3^{6 \cdot 1429}}$.

The purpose of this paper is to demonstrate that, with modest computational resources, the new algorithms can be used to solve instances of the discrete logarithm problem that remain beyond the reach of classical algorithms. The first target field is the 1303-bit field $\mathbb{F}_{3^{6 \cdot 137}}$; this field does not enjoy any Kummer-like properties. More precisely, we are interested in solving the discrete logarithm problem in the order-r subgroup $\mathcal{G}$ of $\mathbb{F}_{3^{6 \cdot 137}}^{*}$, where $r = (3^{137} - 3^{69} + 1)/$ 7011427850892440647 is a 155-bit prime. The discrete logarithm problem in this group is of cryptographic interest because the values of the bilinear pairing derived from the supersingular elliptic curve $E : y^2 = x^3 - x + 1$ over $\mathbb{F}_{3^{137}}$ lie in $\mathcal{G}$.[1] Consequently, if logarithms in $\mathcal{G}$ can be computed efficiently then the associated bilinear pairing is rendered cryptographically insecure. Note that since r is a 155-bit prime, Pollard's rho algorithm [29] for computing logarithms in $\mathcal{G}$ is infeasible. Moreover, recent work on computing logarithms in the 809-bit field $\mathbb{F}_{2^{809}}$ [3] suggests that Coppersmith's algorithm is infeasible for computing logarithms in $\mathcal{G}$, whereas recent work on computing logarithms in the 923-bit field $\mathbb{F}_{3^{6 \cdot 97}}$ [22] (see also [30]) indicates that computing logarithms in $\mathcal{G}$ using the Joux-Lercier algorithm [25] would be a formidable challenge. In contrast, we show that Joux's algorithm can be used to compute logarithms in $\mathcal{G}$ in just a few days using a small number of CPUs; more precisely, our computation consumed a total of 888 CPU hours. The computational effort expended in our experiment is relatively small, despite the fact that our implementation was done using the computer algebra system Magma V2.20-2 [27] and is far from optimal.

The second target field is the 1551-bit field $\mathbb{F}_{3^{6 \cdot 163}}$; this field does not enjoy any Kummer-like properties. More precisely, we are interested in solving the discrete logarithm problem in the order-r subgroup $\mathcal{G}$ of $\mathbb{F}_{3^{6 \cdot 163}}^{*}$, where $r = 3^{163} + 3^{82} + 1$ is a 259-bit prime. The discrete logarithm problem in this group is of cryptographic interest because the values of the bilinear pairing derived from the supersingular elliptic curve $E : y^2 = x^3 - x - 1$ over $\mathbb{F}_{3^{163}}$ lie in $\mathcal{G}$. This bilinear pairing was first considered by Boneh, Lynn and Shacham in their landmark paper on short signature schemes [8]; see also [20]. Furthermore, the bilinear pairing derived from the quadratic twist of E was one of the pairings implemented by Galbraith, Harrison and Soldera [14]. Again, we show that Joux's algorithm can be used to compute logarithms in $\mathcal{G}$ in just a few days using a small number of CPUs; our computation used 1201 CPU hours.

---

[1] We note that the supersingular elliptic curves $y^2 = x^3 - x \pm 1$ over $\mathbb{F}_{3^n}$ have embedding degree 6 and were proposed for cryptographic use in several early papers on pairing-based cryptography [5, 8, 14, 19].

After we had completed the $\mathbb{F}_{3^{6\cdot137}}$ discrete logarithm computation, Granger, Kleinjung and Zumbrägel [18] presented several practical improvements and refinements of Joux's algorithm. These improvements allowed them to compute logarithms in the 4404-bit field $\mathbb{F}_{2^{12\cdot367}}$ in approximately 52,240 CPU hours, and drastically lowered the estimated time to compute logarithms in the 4892-bit field $\mathbb{F}_{2^{4\cdot1223}}$ to 2^{59} modular multiplications. More recently, Joux and Pierrot [26] presented a more efficient algorithm for computing logarithms of factor base elements. The new algorithm was used to compute logarithms in the 3796-bit characteristic-three field $\mathbb{F}_{3^{5\cdot479}}$ in less than 8600 CPU hours.

The remainder of the paper is organized as follows. In Sect. 2, we review Joux's algorithm for computing logarithms in $\mathbb{F}_{q^{3n}}$; the algorithm uses the polynomial representation (selection of h_0 and h_1) of Granger and Zumbrägel [21]. Our experimental results with computing logarithms in $\mathbb{F}_{3^{6\cdot137}}$ and $\mathbb{F}_{3^{6\cdot163}}$ are reported in Sects. 3 and 4, respectively. In Sect. 5, we use the aforementioned improvements from [18] and [26] to derive improved upper bounds for discrete logarithm computations in $\mathbb{F}_{3^{6\cdot509}}$ and $\mathbb{F}_{3^{6\cdot1429}}$. We draw our conclusions in Sect. 6.

2   Joux's $L[1/4 + o(1)]$ Algorithm

Let $\mathbb{F}_{q^{3n}}$ be a finite field where $n \leq 2q+1$.[2] The elements of $\mathbb{F}_{q^{3n}}$ are represented as polynomials of degree at most $n-1$ over $\mathbb{F}_{q^3}$. Let $N = q^{3n} - 1$, and let r be a prime divisor of N. In this paper, we are interested in the discrete logarithm problem in the order-r subgroup of $\mathbb{F}_{q^{3n}}^*$. More precisely, we are given two elements α, β of order r in $\mathbb{F}_{q^{3n}}^*$ and we wish to find $\log_\alpha \beta$. Let g be an element of order N in $\mathbb{F}_{q^{3n}}^*$. Then $\log_\alpha \beta = (\log_g \beta)/(\log_g \alpha) \bmod r$. Thus, in the remainder of this section we will assume that we need to compute $\log_g h \bmod r$, where h is an element of order r in $\mathbb{F}_{q^{3n}}^*$.

The algorithm proceeds by first finding the logarithms (mod r) of all degree-one elements in $\mathbb{F}_{q^{3n}}$ (Sect. 2.2). Then, in the *descent stage*, $\log_g h$ is expressed as a linear combination of logarithms of degree-one elements. The descent stage proceeds in several steps, each expressing the logarithm of a degree-D element as a linear combination of the logarithms of elements of degree $\leq m$ for some $m < D$. Four descent methods are employed; these are described in Sects. 2.3–2.6.

Notation. $N_{q^3}(m, n)$ denotes the number of monic m-smooth degree-n polynomials in $\mathbb{F}_{q^3}[X]$, $A_{q^3}(m, n)$ denotes the average number of distinct monic irreducible factors among all monic m-smooth degree-n polynomials in $\mathbb{F}_{q^3}[X]$, and $S_{q^3}(m, d)$ denotes the cost of testing m-smoothness of a degree-d polynomial in $\mathbb{F}_{q^3}[X]$. Formulas for $N_{q^3}(m, n)$, $A_{q^3}(m, n)$ and $S_{q^3}(m, n)$ are given in [1]. For $\gamma \in \mathbb{F}_{q^3}$, $\overline{\gamma}$ denotes the element γ^{q^2}. For $P \in \mathbb{F}_{q^3}[X]$, $\overline{P}$ denotes the polynomial obtained by raising each coefficient of P to the power q^2. The cost of an integer

---

[2] More generally, one could consider fields $\mathbb{F}_{q^{kn}}$ where $n \leq 2q + 1$. We focus on the case $k = 3$ since our target fields are $\mathbb{F}_{3^{6n}}$ with $n \in \{137, 163\}$, which we will embed in $\mathbb{F}_{(3^4)^{3\cdot n}}$.

addition modulo r is denoted by A_r, and the cost of a multiplication in $\mathbb{F}_{q^3}$ is denoted by M_{q^3}. The projective general linear group of degree 2 over $\mathbb{F}_q$ is denoted $\mathrm{PGL}_2(\mathbb{F}_q)$. $\mathcal{P}_q$ is a set of distinct representatives of the left cosets of $\mathrm{PGL}_2(\mathbb{F}_q)$ in $\mathrm{PGL}_2(\mathbb{F}_{q^3})$; note that $\#\mathcal{P}_q = q^6 + q^4 + q^2$. A matrix $\left(\begin{smallmatrix} a & b \\ c & d \end{smallmatrix}\right) \in \mathcal{P}_q$ is identified with the quadruple (a, b, c, d).

2.1 Setup

Select polynomials $h_0, h_1 \in \mathbb{F}_{q^3}[X]$ of small degree so that

$$X \cdot h_1(X^q) - h_0(X^q) \tag{1}$$

has an irreducible factor I_X of degree n in $\mathbb{F}_{q^3}[X]$; we will henceforth assume that $\max(\deg h_0, \deg h_1) = 2$, whence $n \le 2q + 1$. Note that

$$X \equiv \frac{h_0(X^q)}{h_1(X^q)} \equiv \left(\frac{\overline{h}_0(X)}{\overline{h}_1(X)}\right)^q \pmod{I_X}. \tag{2}$$

The field $\mathbb{F}_{q^{3n}}$ is represented as $\mathbb{F}_{q^{3n}} = \mathbb{F}_{q^3}[X]/(I_X)$ and the elements of $\mathbb{F}_{q^{3n}}$ are represented as polynomials in $\mathbb{F}_{q^3}[X]$ of degree at most $n - 1$. Let g be a generator of $\mathbb{F}_{q^{3n}}^*$.

2.2 Finding Logarithms of Linear Polynomials

Let $\mathcal{B}_1 = \{X + a \mid a \in \mathbb{F}_{q^3}\}$, and note that $\#\mathcal{B}_1 = q^3$. To compute the logarithms of $\mathcal{B}_1$-elements, we first generate linear relations of these logarithms. Let $(a, b, c, d) \in \mathcal{P}_q$. Substituting $Y \mapsto (aX + b)/(cX + d)$ into the systematic equation

$$Y^q - Y = \prod_{\alpha \in \mathbb{F}_q} (Y - \alpha) \tag{3}$$

and using (2) yields

$$\left((aX + b)(\overline{c}\,\overline{h}_0 + \overline{d}\,\overline{h}_1) - (\overline{a}\,\overline{h}_0 + \overline{b}\,\overline{h}_1)(cX + d)\right)^q \tag{4}$$

$$\equiv \overline{h}_1^q \cdot (cX + d) \cdot \prod_{\alpha \in \mathbb{F}_q} [(a - \alpha c)X + (b - \alpha d)].$$

If the polynomial on the left side of (4) is 1-smooth, then taking logarithms (mod r) of both sides of (4) yields a linear relation of the logarithms of $\mathcal{B}_1$-elements and the logarithm of $\overline{h}_1$. The probability that the left side of (4) is 1-smooth is $N_{q^3}(1, 3)/q^9 \approx \frac{1}{6}$. Thus, after approximately $6q^3$ trials one expects to obtain q^3 relations. The cost of the relation generation stage is $6q^3 \cdot S_{q^3}(1, 3)$. The logarithms can then be obtained by using Wiedemann's algorithm for solving sparse systems of linear equations [10,31]. The expected cost of the linear algebra is $q^7 \cdot A_r$ since each equation has approximately q nonzero terms.

2.3 Continued-Fractions Descent

Recall that we wish to compute $\log_g h \bmod r$, where $h \in \mathbb{F}_{q^{3n}} = \mathbb{F}_{q^3}[X]/(I_X)$ has order r. We will henceforth assume that $\deg h = n - 1$. The descent stage begins by multiplying h by a random power of g. The extended Euclidean algorithm is used to express the resulting field element h' in the form $h' = w_1/w_2$ where $\deg w_1, \deg w_2 \approx n/2$ [7]; for simplicity, we shall assume that n is odd and $\deg w_1 = \deg w_2 = (n-1)/2$. This process is repeated until both w_1 and w_2 are m-smooth for some chosen $m < (n-1)/2$. This gives $\log_g h'$ as a linear combination of logarithms of polynomials of degree at most m. The expected cost of this continued-fractions descent step is approximately

$$\left(\frac{(q^3)^{(n-1)/2}}{N_{q^3}(m, (n-1)/2)} \right)^2 \cdot S_{q^3}(m, (n-1)/2). \tag{5}$$

The expected number of distinct irreducible factors of w_1 and w_2 is $2A_{q^3}(m, (n-1)/2)$. In the concrete analysis, we shall assume that each of these irreducible factors has degree exactly m. The logarithm of each of these degree-m polynomials is then expressed as a linear combination of logarithms of smaller degree polynomials using one of the descent methods described in Sects. 2.4, 2.5 and 2.6.

2.4 Classical Descent

Let p be the characteristic of $\mathbb{F}_q$, and let $q = p^\ell$. Let $s \in [0, \ell]$, and let $R \in \mathbb{F}_{q^3}[X, Y]$. Then it can be seen that

$$\left[R(X, (\overline{h}_0/\overline{h}_1)^{p^{\ell-s}}) \right]^{p^s} \equiv R'(X^{p^s}, X) \pmod{I_X} \tag{6}$$

where R' is obtained from R by raising all its coefficients to the power p^s. Let $\mu = \deg_Y R$. Then multiplying both sides of (6) by $\overline{h}_1^{q\mu}$ gives

$$\left[\overline{h}_1^{p^{\ell-s} \cdot \mu} \cdot R(X, (\overline{h}_0/\overline{h}_1)^{p^{\ell-s}}) \right]^{p^s} \equiv \overline{h}_1^{q\mu} \cdot R'(X^{p^s}, X) \pmod{I_X}. \tag{7}$$

Let $Q \in \mathbb{F}_{q^3}[X]$ with $\deg Q = D$, and let $m < D$. In the Joux-Lercier descent method [25], as modified by Gölöğlu et al. [15], one selects $s \in [0, \ell]$ and searches for a polynomial $R \in \mathbb{F}_{q^3}[X, Y]$ such that (i) $Q \mid R_2$ where $R_2 = R'(X^{p^s}, X)$; (ii) $\deg R_1$ and $\deg R_2/Q$ are appropriately balanced where $R_1 = \overline{h}_1^{p^{\ell-s}\mu} R(X, (\overline{h}_0/\overline{h}_1)^{p^{\ell-s}})$; and (iii) both R_1 and R_2/Q are m-smooth. Taking logarithms of both sides of (7) then gives an expression for $\log_g Q$ in terms of the logarithms of polynomials of degree at most m.

A family of polynomials R satisfying (i) and (ii) can be constructed by finding a basis $\{(u_1, u_2), (v_1, v_2)\}$ of the lattice

$$L_Q = \{(w_1, w_2) \in \mathbb{F}_{q^3}[X] \times \mathbb{F}_{q^3}[X] : Q \mid (w_1(X) - w_2(X)X^{p^s})\}$$

where $\deg u_1$, $\deg u_2$, $\deg v_1$, $\deg v_2 \approx D/2$. By writing $(w_1, w_2) = a(u_1, u_2) + b(v_1, v_2) = (au_1 + bv_1, au_2 + bv_2)$ with $a \in \mathbb{F}_{q^3}[X]$ monic of degree δ and $b \in \mathbb{F}_{q^3}[X]$ of degree $\delta - 1$, the points (w_1, w_2) in L_Q can be sampled to obtain polynomials $R(X, Y) = w_1''(Y) - w_2''(Y)X$ satisfying (i) and (ii) where w'' is obtained from w by raising all its coefficients to the power p^{-s}. The number of lattice points to consider is therefore $(q^3)^{2\delta}$. We have $\deg w_1, \deg w_2 \approx D/2 + \delta$, so $\deg R_1 = t_1 \approx 2(D/2 + \delta)p^{\ell-s} + 1$ and $\deg R_2 = t_2 \approx (D/2 + \delta) + p^s$. In order to ensure that there are sufficiently many such lattice points to generate a polynomial R for which both R_1 and R_2/Q are m-smooth, the parameters s and δ must be selected so that

$$q^{6\delta} \gg \frac{q^{3t_1}}{N_{q^3}(m, t_1)} \cdot \frac{q^{3(t_2 - D)}}{N_{q^3}(m, t_2 - D)}. \tag{8}$$

Ignoring the time to compute a balanced basis of L_Q, the expected cost of finding a polynomial R satisfying (i)–(iii) is

$$\frac{q^{3t_1}}{N_{q^3}(m, t_1)} \cdot \frac{q^{3(t_2 - D)}}{N_{q^3}(m, t_2 - D)} \cdot \min(S_{q^3}(m, t_1), S_{q^3}(m, t_2 - D)). \tag{9}$$

The expected number of distinct irreducible factors of R_1 and R_2/Q is $A_{q^3}(m, t_1) + A_{q^3}(m, t_2 - D)$.

2.5 Gröbner bases descent

Let $Q \in \mathbb{F}_{q^3}[X]$ with $\deg Q = D$. Let $m = \lceil (D+1)/2 \rceil$, and suppose that $3m < n$. In Joux's new descent method [23, Sect. 5.3], one finds degree-m polynomials $k_1, k_2 \in \mathbb{F}_{q^3}[X]$ such that $G = k_1 \tilde{k}_2 - \tilde{k}_1 k_2 = QR$, where $\tilde{k}_1 = \overline{h}_1^m \overline{k}_1(\overline{h}_0/\overline{h}_1)$ and $\tilde{k}_2 = \overline{h}_1^m \overline{k}_2(\overline{h}_0/\overline{h}_1)$, and $R \in \mathbb{F}_{q^3}[X]$. Note that $\deg R = 3m - D$. If R is m-smooth, then we obtain a linear relationship between $\log_g Q$ and logs of degree-m polynomials (see [2, Sect. 3.7]):

$$\overline{h}_1^{-mq} \cdot k_2 \cdot \prod_{\alpha \in \mathbb{F}_q} (k_1 - \alpha k_2) \equiv (Q(X)R(X))^q \pmod{I_X}. \tag{10}$$

To determine (k_1, k_2, R) that satisfy

$$k_1 \tilde{k}_2 - \tilde{k}_1 k_2 = QR, \tag{11}$$

one can transform (11) into a system of multivariate quadratic equations over $\mathbb{F}_q$. Specifically, each coefficient of k_1, k_2 and R is written using three variables over $\mathbb{F}_q$. The coefficients of $\tilde{k}_1$ and $\tilde{k}_2$ can then be written in terms of the coefficients of k_1 and k_2. Hence, equating coefficients of X^i of both sides of (11) yields $3m + 1$ quadratic equations. Equating $\mathbb{F}_q$-components of these equations then yields $9m + 3$ bilinear equations in $15m - 3D + 9$ variables over $\mathbb{F}_q$. This system of equations can be solved by finding a Gröbner basis for the ideal it generates. Finally, solutions (k_1, k_2, R) are tested until one is found for which

R is m-smooth. This yields an expression for $\log_g Q$ in terms of the logarithms of approximately $q + 1 + A_{q^3}(m, 3m - D)$ polynomials of degree (at most) m; in the concrete analysis we shall assume that each of the polynomials has degree exactly m.

2.6 2-to-1 Descent

The Gröbner bases descent methodology of Sect. 2.5 can be employed in the case $(D, m) = (2, 1)$. However, as also reported by Joux in his $\mathbb{F}_{2^{6168}}$ discrete log computation [24], we found the descent to be successful for only about 50 % of all irreducible quadratic polynomials. Despite this, some strategies can be used to increase this percentage.

Let $Q(X) = X^2 + uX + v \in \mathbb{F}_{q^3}[X]$ be an irreducible quadratic polynomial for which the Gröbner bases descent method failed.

Strategy 1. Introduced by Joux [24] and Göloğlu et al. [16], this strategy is based on the systematic equation derived from $Y^{q'} - Y$ where $q' < q$ and $\mathbb{F}_{q'}$ is a proper subfield of $\mathbb{F}_{q^3}$ instead of the systematic Eq. (3) derived from $Y^q - Y$. Let p be the characteristic of $\mathbb{F}_q$, and let $q = p^\ell$, $q' = p^{\ell'}$, and $s = \ell - \ell'$. Then $q = p^s \cdot q'$. Now, one searches for $a, b, c, d \in \mathbb{F}_{q^3}$ such that

$$G = (aX + b)(\overline{c}\overline{h}_0 + \overline{d}\,\overline{h}_1)^{p^s} - (\overline{a}\overline{h}_0 + \overline{b}\,\overline{h}_1)^{p^s}(cX + d) = QR$$

with $R \in \mathbb{F}_{q^3}[X]$. Note that $\deg R = 2p^s - 1$.[3] If R is 1-smooth, then we obtain a linear relationship between $\log_g Q$ and logs of linear polynomials since

$$G^q \equiv \overline{h}_1^{p^s q} \cdot (cX + d)^{p^s} \cdot \prod_{\alpha \in \mathbb{F}_{q'}} \left((aX + b)^{p^s} - \alpha (cX + d)^{p^s} \right) \pmod{I_X},$$

as can be seen by making the substitution $Y \mapsto (aX + b)^{p^s}/(cX + d)^{p^s}$ into the systematic equation derived from $Y^{q'} - Y$.

Unfortunately, in all instances we considered, the polynomial R never factors completely into linear polynomials. However, it hopefully factors into a quadratic polynomial Q' and $2p^s - 3$ linear polynomials, thereby yielding a relation between Q and another quadratic which has a roughly 50 % chance of descending using Gröbner bases descent. Combined with the latter, this strategy descends about 95 % of all irreducible quadratic polynomials in the fields $\mathbb{F}_{3^{6 \cdot 137}}$ and $\mathbb{F}_{3^{6 \cdot 163}}$.

Strategy 2. We have

$$\overline{h}_1^{2q} Q(X) \equiv \overline{h}_1^{2q} Q((\overline{h}_0/\overline{h}_1)^q) = \overline{h}_0^{2q} + u\overline{h}_0^q \overline{h}_1^q + v\overline{h}_1^{2q}$$
$$= (\overline{h}_0^2 + \overline{u}\overline{h}_0 \overline{h}_1 + \overline{v}\,\overline{h}_1^2)^q \pmod{I_X}. \qquad (12)$$

---

[3] For our $\mathbb{F}_{3^{6 \cdot 137}}$ and $\mathbb{F}_{3^{6 \cdot 163}}$ computations, we have $q = 3^4$ and used $q' = 3^3$, so $s = 1$ and $\deg R = 5$.

It can be seen that the degree-4 polynomial $f_Q(X) = \overline{h}_0^2 + \overline{u}\overline{h}_0\overline{h}_1 + \overline{v}\,\overline{h}_1^2$ is either a product of two irreducible quadratics or itself irreducible. In the former case, we apply the standard Gröbner bases descent method to the two irreducible quadratics. If both descents are successful, then we have succeeded in descending the original Q.

The strategies are combined in the following manner. For an irreducible quadratic $Q \in \mathbb{F}_{q^3}[X]$, we first check if the Gröbner bases descent is successful. If the descent fails, we apply Strategy 2 to Q. In the case where f_Q factors into two irreducible quadratics, and at least one of them fails to descend with Gröbner bases descent, we apply Strategy 1 to Q. If Strategy 1 fails on Q, we apply it to the two quadratic factors of f_Q. In the case where f_Q is irreducible, we apply Strategy 1 to Q.

If none of the attempts succeed, we declare Q to be "bad", and avoid it in the higher-degree descent steps by repeating a step until all the quadratics encountered are "good". In our experiments with $\mathbb{F}_{3^{6\cdot137}}$ and $\mathbb{F}_{3^{6\cdot163}}$, we observed that approximately 97.2 % of all irreducible quadratic polynomials Q were "good".

To see that this percentage is sufficient to complete the descent phase in these two fields, consider a 3-to-2 descent step where the number of resulting irreducible quadratic polynomials is 42 on average (cf. Eq. (10)). Then the probability of descending a degree-3 polynomial after finding one useful solution (k_1, k_2, R) in Gröbner bases descent is $0.972^{42} \approx 0.3$. Therefore, after at most four trials we expect to successfully descend a degree-3 polynomial. Since the expected number of distinct solutions of (11) is approximately q^3 (according to Eq. (10) in [18]), one can afford this many trials.

3   Computing Discrete Logarithms in $\mathbb{F}_{3^{6\cdot137}}$

The supersingular elliptic curve $E : y^2 = x^3 - x + 1$ has order $\#E(\mathbb{F}_{3^{137}}) = cr$, where

$$c = 7 \cdot 4111 \cdot 5729341 \cdot 42526171$$

and

$$r = (3^{137} - 3^{69} + 1)/c = 3309828011909019102877558005508217505642849\,5623$$

is a 155-bit prime. The Weil and Tate pairing attacks [13,28] efficiently reduce the discrete logarithm problem in the order-r subgroup $\mathcal{E}$ of $E(\mathbb{F}_{3^{137}})$ to the discrete logarithm problem in the order-r subgroup $\mathcal{G}$ of $\mathbb{F}_{3^{6\cdot137}}^*$.

Our approach to computing logarithms in $\mathcal{G}$ is to use Joux's algorithm to compute logarithms in the quadratic extension $\mathbb{F}_{3^{12\cdot137}}$ of $\mathbb{F}_{3^{6\cdot137}}$ (so $q = 3^4$ and $n = 137$ in the notation of Sect. 2). More precisely, we are given two elements α, β of order r in $\mathbb{F}_{3^{12\cdot137}}^*$ and we wish to find $\log_\alpha \beta$. Let g be a generator of $\mathbb{F}_{3^{12\cdot137}}^*$. Then $\log_\alpha \beta = (\log_g \beta)/(\log_g \alpha) \bmod r$. Thus, in the remainder of the section we will assume that we need to compute $\log_g h \bmod r$, where h is an element of order r in $\mathbb{F}_{3^{12\cdot137}}^*$.

The DLP instance we solved is described in Sect. 3.1. The concrete estimates from Sect. 2 for solving the DLP instances are given in Sect. 3.2. These estimates are only upper bounds on the running time of the algorithm. Nevertheless, they provide convincing evidence for the feasibility of the discrete logarithm computations. Our experimental results are presented in Sect. 3.3.

3.1 Problem Instance

Let N denote the order of $\mathbb{F}^*_{3^{12\cdot137}}$. Using the tables from the Cunningham Project [11], we determined that the factorization of N is $N = p_1^4 \cdot \prod_{i=2}^{31} p_i$, where the p_i are the following primes (and $r = p_{25}$):

$p_1 = 2$ $p_2 = 5$ $p_3 = 7$ $p_4 = 13$ $p_5 = 73$ $p_6 = 823$ $p_7 = 4111$ $p_8 = 4933$

$p_9 = 236737$ $p_{10} = 344693$ $p_{11} = 2115829$ $p_{12} = 5729341$ $p_{13} = 42526171$

$p_{14} = 217629707$ $p_{15} = 634432753$ $p_{16} = 685934341$ $p_{17} = 82093596209179$

$p_{18} = 4354414202063707$ $p_{19} = 18329390240606021$ $p_{20} = 46249052722878623693$

$p_{21} = 201820452878622271249$ $p_{22} = 1139388291348802249541428925264777$

$p_{23} = 518545466463281867910174177004304863965137$

$p_{24} = 273537065683369412556889640428278023763717$

$p_{25} = 330982801190901910287755800550821750564284956237$

$p_{26} = 70671225820194025466782664267300876838722911504837937$

$p_{27} = 10808180977383999518825680049914154368439303545035055137$

$p_{28} = 91321974595662761339222271626247966116126450162880692588587183952237$

$p_{29} = 39487531149773489532096996293368370182957526257988573877031054477249$
      393549

$p_{30} = 40189860022384850044254854796561182547553072730738823866986300807613$
      29207749418522920289

$p_{31} = 190643231538252720728036858708039556228342865231390374035807523108227$
      $8966446469840637369426240662274068981321133662265931584644197137$.

We chose $\mathbb{F}_{3^4} = \mathbb{F}_3[U]/(U^4 + U^2 + 2)$ and $\mathbb{F}_{3^{12}} = \mathbb{F}_{3^4}[V]/(V^3 + V + U^2 + U)$, and selected $h_0(X) = V^{326196}X^2 + V^{35305}X + V^{204091} \in \mathbb{F}_{3^{12}}[X]$ and $h_1 = 1$. Then $I_X \in \mathbb{F}_{3^{12}}[X]$ is the degree-137 monic irreducible factor of $X - h_0(X^{3^4})$; the other irreducible factor has degree 25.

We chose the generator $g = X + V^{113713}$ of $\mathbb{F}^*_{3^{12\cdot137}}$. To generate an order-$r$ discrete logarithm challenge h, we computed

$$h' = \sum_{i=0}^{136} \left(V^{\lfloor \pi \cdot (3^{12})^{i+1} \rfloor \bmod 3^{12}} \right) X^i$$

and then set $h = (h')^{N/r}$. The discrete logarithm $\log_g h \bmod r$ was found to be

$$x = 273396190769750939202455159732141869630256565 59.$$

This can be verified by checking that $h = (g^{N/r})^y$, where $y = x \cdot (N/r)^{-1} \bmod r$ (cf. Appendix A).

3.2   Estimates

The factor base $\mathcal{B}_1$ has size $3^{12} \approx 2^{19}$. The cost of the relation generation is approximately $2^{29.2} M_{q^3}$, whereas the cost of the linear algebra is approximately $2^{44.4} A_r$. Figure 1 shows the estimated running times for the descent stage. Further information about the parameter choices are provided below.

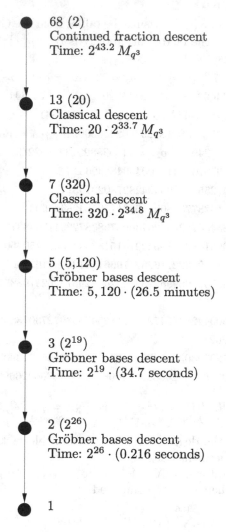

68 (2)
Continued fraction descent
Time: $2^{43.2} M_{q^3}$

13 (20)
Classical descent
Time: $20 \cdot 2^{33.7} M_{q^3}$

7 (320)
Classical descent
Time: $320 \cdot 2^{34.8} M_{q^3}$

5 (5,120)
Gröbner bases descent
Time: $5,120 \cdot (26.5 \text{ minutes})$

3 (2^{19})
Gröbner bases descent
Time: $2^{19} \cdot (34.7 \text{ seconds})$

2 (2^{26})
Gröbner bases descent
Time: $2^{26} \cdot (0.216 \text{ seconds})$

1

Fig. 1. A typical path of the descent tree for computing an individual logarithm in $\mathbb{F}_{3^{12 \cdot 137}}$ ($q = 3^4$). The numbers in parentheses next to each node are the expected number of nodes at that level. 'Time' is the expected time to generate all nodes at a level.

1. For the continued-fractions descent stage, we selected $m = 13$. The expected cost of this descent is $2^{43.2} M_{q^3}$, and the expected number of irreducible factors of degree (at most) 13 obtained is $2A_{3^{12}}(68, 13) \approx 20$.

2. Two classical descent stages are employed. In the first stage, we have $D = 13$ and select $m = 7$, $s = 3$, $\delta = 1$, which yield $t_1 = 43$ and $t_2 = 34$. The expected cost of the descent for each of the 20 degree-13 polynomials is approximately $2^{33.7} M_{q^3}$. The expected total number of distinct irreducible polynomials of degree (at most) 7 obtained is approximately 320.

   In the second classical descent stage, we have $D = 7$ and select $m = 5$, $s = 3$, $\delta = 1$, which yield $t_1 = 25$ and $t_2 = 31$. The expected cost of the descent for each of the 320 degree-7 polynomials is approximately $2^{34.8} M_{q^3}$. The expected total number of distinct irreducible polynomials of degree (at most) 5 obtained is approximately 5,120.

3. Our implementation of the Gröbner bases descent stage used Magma's implementation of Faugére's F4 algorithm [12] and took 26.5 min on average for a 5-to-3 descent, 34.7 s for a 3-to-2 descent, and 0.216 s for a 2-to-1 descent. The total expected running time for each of these stages is 94, 211 and 168 days, respectively.

Since all the descent stages can be effectively parallelized, our estimates suggest that a discrete logarithm can be computed in a week or so given a few dozen processors. In fact (and as confirmed by our experimental results), the actual running time is expected to be significantly less than the estimated running time since the estimates are quite conservative; for example, our estimates for the number of branches in a descent step assumes that each distinct irreducible polynomial has degree exactly m, whereas in practice many of these polynomials will have degree significantly less than m.

3.3 Experimental Results

Our experiments were run on an Intel i7-2600K 3.40 GHz machine (Sandy Bridge), and on an Intel i7-4700MQ 2.40 GHz machine (Haswell).

Relation generation took 1.05 CPU hours (Sandy Bridge, 1 core). The resulting sparse linear system of linear equation was solved using Magma's multi-threaded parallel version of the Lanczos algorithm; the computation took 556.8 CPU hours (Sandy Bridge, 4 cores).

In the continued-fractions descent stage, the first degree-68 polynomial yielded 9 irreducible factors of degrees 12, 12, 11, 10, 8, 6, 6, 2, 1, and the second degree-68 polynomial yielded 11 irreducible factors of degrees 13, 12, 10, 10, 7, 6, 5, 2, 1, 1, 1. The computation took 22 CPU hours (Haswell, 4 cores).

Classical descent was used on the 9 polynomials of degree ≥ 8 to obtain polynomials of degree ≤ 7, and then on the 23 polynomials of degree 7 and 23 polynomials of degree 6 to obtain polynomials of degree ≤ 5. These computations took 80 CPU hours (Haswell, 4 cores).

Finally, we used 5-to-3, 4-to-3, 3-to-2 and 2-to-1 Gröbner bases descent procedures. The average time for a 4-to-3 descent was 33.8 s; the other average times are given in Fig. 1. In total, we performed 233 5-to-3 descents, 174 4-to-3 descents, and 11573 3-to-2 descents. These computations took 115.2 CPU hours, 1.5 CPU hours, and 111.2 CPU hours, respectively (Haswell, 4 cores). We also performed 493537 2-to-1 descents; their running times are incorporated into the running times for the higher-level descents.

4   Computing Discrete Logarithms in $\mathbb{F}_{3^{6 \cdot 163}}$

The supersingular elliptic curve $E : y^2 = x^3 - x - 1$ has order $\#E(\mathbb{F}_{3^{163}}) = 3^{163} + 3^{82} + 1 = r$, where r is the following 259-bit prime:

$$r = 58988115142665874085422772558073634885064063229737341409179099550575662\,3268837.$$

The Weil and Tate pairing attacks [13,28] efficiently reduce the discrete logarithm problem in the order-r group $\mathcal{E} = E(\mathbb{F}_{3^{163}})$ to the discrete logarithm problem in the order-r subgroup $\mathcal{G}$ of $\mathbb{F}^*_{3^{6 \cdot 163}}$.

As in Sect. 3, we will compute logarithms in $\mathcal{G}$ by using Joux's algorithm to compute logarithms in the quadratic extension $\mathbb{F}_{3^{12 \cdot 163}}$ of $\mathbb{F}_{3^{6 \cdot 163}}$ (so $q = 3^4$ and $n = 163$ in the notation of Sect. 2). We will compute $\log_g h \bmod r$, where g is a generator of $\mathbb{F}^*_{3^{12 \cdot 163}}$ and h is an element of order r in $\mathbb{F}^*_{3^{12 \cdot 163}}$.

4.1   Problem Instance

Let N denote the order of $\mathbb{F}^*_{3^{12 \cdot 163}}$. Using the tables from the Cunningham Project [11], we partially factored N as $N = C \cdot p_1^4 \cdot \prod_{i=2}^{22} p_i$, where the p_i are the following primes (and $r = p_{20}$):

$p_1 = 2$ $p_2 = 5$ $p_3 = 7$ $p_4 = 13$ $p_5 = 73$ $p_6 = 653$ $p_7 = 50857$

$p_8 = 107581$ $p_9 = 489001$ $p_{10} = 105451873$ $p_{11} = 380998157$

$p_{12} = 8483499631$ $p_{13} = 5227348213873$ $p_{14} = 8882811705390167$

$p_{15} = 4956470591980320134353$ $p_{16} = 23210817035829275705929$

$p_{17} = 3507171060957186767994912136200333814689659449$

$p_{18} = 6351885141964057411259499526611848626072045955243$

$p_{19} = 84268735918094105836318246511533764121140010481130741067443071103148\,817701717$

$p_{20} = 58988115142665874085422772558073634885064063229737341409179099550575\,6623268837$

$p_{21} = 13262905784043723370034025667618121081540438283177268680045186884853\,26204127242781054287716913828905695771535319617625904849821802388801$

$p_{22} = 24879984727675011205198718183055547601122582974374576908898869641570\,09269122423985704395925964922959410448009886539842494955927136450643\,31019158574269,$

and C is the following 919-bit composite number

$C = 2873322036656120507394501949912283436722983546265951551507632957325767$
$\qquad 0275216328747773792566523729655097848102113488795698936768394494992621$
$\qquad 2312022819011019340957620502000045691081669475648919901346991751981450$
$\qquad 8311534570945558522228827298337826215043744094861514754454151493177.$

We verified that $\gcd(C, N/C) = 1$ and that C is not divisible by any of the first 10^7 primes. Consequently, if an element g is selected uniformly at random from $\mathbb{F}_{3^{12\cdot163}}^*$, and g satisfies $g^{(N-1)/p_i} \neq 1$ for $1 \leq i \leq 22$, then g is a generator with very high probability.[4]

We chose $\mathbb{F}_{3^4} = \mathbb{F}_3[U]/(U^4 + U^2 + 2)$ and $\mathbb{F}_{3^{12}} = \mathbb{F}_{3^4}[V]/(V^3 + V + U^2 + U)$, and selected $h_0(X) = 1$ and

$$h_1(X) = X^2 + V^{530855} \in \mathbb{F}_{3^{12}}[X].$$

Then $I_X \in \mathbb{F}_{3^{12}}[X]$ is the degree-163 irreducible polynomial $X \cdot h_1(X^{3^4}) - 1$:

$$I_X = X^{163} + V^{530855}X + 2.$$

We chose $g = X + V^2$, which we hope is a generator of $\mathbb{F}_{3^{12\cdot163}}^*$.

To generate an order-r discrete logarithm challenge h, we computed

$$h' = \sum_{i=0}^{162} \left(V^{\lfloor \pi \cdot (3^{12})^{i+1} \rfloor \bmod 3^{12}} \right) X^i$$

and then set $h = (h')^{N/r}$. The discrete logarithm $\log_g h \bmod r$ was found to be

$x = 4263959514982791937132913919534490007325925542511325256720397843 5605$
$\qquad 4526194343.$

This can be verified by checking that $h = (g^{N/r})^y$, where $y = x \cdot (N/r)^{-1} \bmod r$ (cf. Appendix B).

4.2   Experimental Results

Our experiments were run on an Intel i7-2600K 3.40 GHz machine (Sandy Bridge), and on an Intel Xeon E5-2650 2.00 GHz machine (Sandy Bridge-EP). The descent strategy was similar to the one used for the $\mathbb{F}_{3^{6\cdot137}}$ computation.

Relation generation took 0.84 CPU hours (Sandy Bridge, 1 core). The resulting sparse system of linear equations was solved using Magma's multi-threaded parallel version of the Lanczos algorithm; the computation took 852.5 CPU hours (Sandy Bridge, 4 cores).

In the continued-fractions descent stage with $m = 15$, the first degree-81 polynomial yielded 8 irreducible factors of degrees 15, 15, 14, 14, 10, 7, 5, 1,

---

[4] More precisely, since C has at most 34 prime factors, each of which is greater than the ten-millionth prime $p = 179424673$, the probability that g is a generator is at least $(1 - \frac{1}{p})^{34} > 0.99999981$.

and the second degree-81 polynomial yielded 12 irreducible factors of degrees 12, 10, 9, 9, 9, 8, 6, 6, 6, 4, 1, 1. The computation took 226.7 CPU hours (Sandy Bridge-EP, 16 cores).

Classical descent was used on the 11 polynomials of degree ≥ 8 to obtain polynomials of degree ≤ 7, and then a variant of classical descent (called the "alternative" method in sect. 3.5 of [2]) was used on the 15 polynomials of degree 7 and 30 polynomials of degree 6 to obtain polynomials of degree ≤ 5. These computations took 51.0 CPU hours (Sandy Bridge-EP, 16 cores).

Finally, we used 5-to-3, 4-to-3 and 3-to-2 Gröbner bases descent procedures. The descent was sped up by writing the coefficients of R (cf. Eq. (11)) in terms of the coefficients of k_1 and k_2; this reduced the number of variables in the resulting bilinear equations from $15m - 3D + 9$ to $9m + 3$. In total, we performed 213 5-to-3 descents, 187 4-to-3 descents, and 11442 3-to-2 descents. These computations took 24.0 CPU hours (Sandy Bridge-EP 16 cores), 0.8 CPU hours (Sandy Bridge, 4 cores), and 44.8 CPU hours (Sandy Bridge, 4 cores), respectively. The running times of the 2-to-1 descents were incorporated into the running times for the higher-level descents.

5   Higher Extension Degrees

As mentioned in Sect. 1, there have been several practical improvements and refinements in discrete logarithm algorithms since Joux's $L[\frac{1}{4} + o(1)]$ algorithm. Most notably, Granger, Kleinjung and Zumbrägel [18] presented several refinements that allowed them to compute logarithms in the 4404-bit characteristic-two field $\mathbb{F}_{2^{12 \cdot 367}}$, and Joux and Pierrot [26] presented a faster algorithm for computing logarithms of factor base elements and used it to compute logarithms in the 3796-bit characteristic-three field $\mathbb{F}_{3^{5 \cdot 479}}$.

In Sect. 5.1, we show that the techniques from [26] and [18] can be used to lower the estimate from [1] for computing discrete logarithms in the 4841-bit characteristic-three field $\mathbb{F}_{3^{6 \cdot 509}}$ from $2^{81.7} M_{q^2}$ to $2^{58.9} M_q$ (where $q = 3^6$). In Sect. 5.2, we use techniques from [18] to lower the estimate from [2] for computing discrete logarithms in the 13590-bit characteristic-three field $\mathbb{F}_{3^{6 \cdot 1429}}$ from $2^{95.8} M_{q^2}$ to $2^{78.8} M_{q^2}$ (where $q = 3^6$). We emphasize that these estimates are *upper bounds* on the running times of known algorithms for computing discrete logarithms. Of course, it is possible that these upper bounds can be lowered with a more judicious choice of algorithm parameters, or with a tighter analysis, or with improvements to the algorithms themselves.

5.1   Computing Discrete Logarithms in $\mathbb{F}_{3^{6 \cdot 509}}$

As in Sect. 4 of [1], we are interested in computing discrete logarithms in the order r-subgroup of $\mathbb{F}_{3^{6 \cdot 509}}^*$, where $r = (3^{509} - 3^{255} + 1)/7$ is an 804-bit prime.

We use the algorithm developed by Joux and Pierrot [26], whence $q = 3^6$ and $k = 1$. The field $\mathbb{F}_{3^6}$ is represented as $\mathbb{F}_3[u]/(u^6 + 2u^4 + u^2 + 2u + 2)$. The field $\mathbb{F}_{3^{6 \cdot 509}}$ is represented as $\mathbb{F}_{3^6}[X]/(I_X)$, where I_X is the degree-509 irreducible factor of $h_1(X)X^q - h_0(X)$ with $h_0(X) = u^{46}X + u^{219}$ and $h_1(X) = X(X + u^{409})$. Joux and Pierrot [26] exploit the special form of $h_0(X)$ and $h_1(X)$ to accelerate the computation of logarithms of polynomials of degree ≤ 4; the dominant step is the computation of logarithms of degree-3 polynomials, where q linear algebra problems are solved each taking time approximately $q^5/27\,A_r$. The continued-fractions, classical and Gröbner bases descents are all performed over $\mathbb{F}_q$.

The new cost estimates are presented in Table 1. We used the estimates for smoothness testing from [17], and the 'bottom-top' approach from [18] for estimating the cost of Gröbner bases descent from degree 15 to degree 4. We assume that 2^{27} multiplications in $\mathbb{F}_{3^6}$ can be performed in 1 s; we achieved this performance using a look-up table approach. The timings for Gröbner bases descent and $\mathbb{F}_{3^6}$ multiplications were obtained on an Intel i7-3930K 3.2 GHz machine. In a non-optimized C implementation, we have observed an A_r cost of 43 clock cycles, where lazy reduction is used to amortize the cost of a modular reduction among many integer additions. This yields the cost ratio $A_r/M_q \approx 2$.

The main effect of the improvements is the removal of the QPA descent stage from the estimates in [1]. The overall running time is $2^{58.9}M_q$, a significant improvement over the $2^{81.7}M_{q^2}$ estimate from [1]. In particular, assuming the availability of processors that can perform 2^{27} $\mathbb{F}_{3^6}$-multiplications per second, the estimated running time is approximately 127 CPU years — this is a feasible computation if one has access to a few hundred cores.

Table 1. Estimated costs of the main steps for computing discrete logarithms in $\mathbb{F}_{3^{6 \cdot 509}}$ ($q = 3^6$). A_r and M_q denote the costs of an addition modulo the 804-bit prime $r = (3^{509} - 3^{255} + 1)/7$ and a multiplication in $\mathbb{F}_{3^6}$. We use the cost ratio $A_r/M_q = 2$, and also assume that 2^{27} multiplications in $\mathbb{F}_{3^6}$ can be performed in 1 s.

Finding logarithms of polynomials of degree ≤ 4		
Linear algebra	$2^{52.3}A_r$	$2^{53.3}M_q$
Descent		
Continued-fractions (254 to 40)	$2^{56.9}M_q$	$2^{56.9}M_q$
Classical (40 to 21)	$12.7 \times 2^{54.2}M_q$	$2^{57.9}M_q$
Classical (21 to 15)	$159 \times 2^{49.4}M_q$	$2^{56.7}M_q$
Gröbner bases (15 to 4)	$1924 \times 8249\text{s}$	$2^{50.9}M_q$

Remark 1. The strategy for computing logarithms in $\mathbb{F}_{3^{6 \cdot 509}}$ can be employed to compute logarithms in $\mathbb{F}_{3^{6 \cdot 239}}$. The latter problem is of cryptographic interest because the prime-order elliptic curve $y^2 = x^3 - x - 1$ over $\mathbb{F}_{3^{239}}$ has embedding degree 6 and has been considered in several papers including [20] and [6]. One could use continued-fractions descent from degree 119 to degree 20 with an estimated cost of $2^{50}M_q$, followed by a classical descent stage from degree 20 to degree 15 at a cost of $2^{53.2}M_q$, and finally Gröbner bases descent to degree 4 at a cost of $2^{47.2}M_q$. The total computational effort is $2^{54.3}M_q$, or approximately 5.2 CPU years.

5.2   Computing Discrete Logarithms in $\mathbb{F}_{3^{6\cdot1429}}$

As in Sect. 4 of [2], we are interested in computing discrete logarithms in the order r-subgroup of $\mathbb{F}_{3^{6\cdot1429}}^*$, where $r = (3^{1429} - 3^{715} + 1)/7622150170693$ is a 2223-bit prime. To accomplish this, we embed $\mathbb{F}_{3^{6\cdot1429}}$ in its quadratic extension $\mathbb{F}_{3^{12\cdot1429}}$. Let $q = 3^6$ and $k = 2$. The field $\mathbb{F}_{3^{12\cdot1429}}$ is represented as $\mathbb{F}_{q^2}[X]/(I_X)$, where I_X is a monic degree-1429 irreducible factor of $h_1(X^q) \cdot X - h_0(X^q)$ with $h_0, h_1 \in \mathbb{F}_{q^2}[X]$ and $\max(\deg h_0, \deg h_1) = 2$.

The techniques from [18] employed to improve the estimates of [2] are the following:

1. Since logarithms are actually sought in the field $\mathbb{F}_{3^{6\cdot1429}}$, the continued fractions and classical descent stages are performed over $\mathbb{F}_q$ (and not $\mathbb{F}_{q^2}$).
2. In the final classical descent stage to degree 11, one permits irreducible factors over $\mathbb{F}_q$ of even degree up to 22; any factors of degree $2t \geq 12$ that are obtained can be written as a product of two degree-t irreducible polynomials over $\mathbb{F}_{q^2}$.
3. The number of irreducible factors of an m-smooth degree-t polynomial is estimated as t/m.
4. The smoothness testing estimates from Appendix B of [17] were used.

The remaining steps of the algorithm, namely finding logarithms of linear polynomial, finding logarithms of irreducible quadratic polynomials, QPA descent, and Gröbner bases descent, are as described in [2].

Table 2. Estimated costs of the main steps for computing discrete logarithms in $\mathbb{F}_{3^{12\cdot1429}}$ ($q = 3^6$). A_r, M_q, and M_{q^2} denote the costs of an addition modulo the 2223-bit prime r, a multiplication in $\mathbb{F}_{3^6}$, and a multiplication in $\mathbb{F}_{3^{12}}$. We use the cost ratio $A_r/M_{q^2} = 4$, and also assume that 2^{26} (resp. 2^{27}) multiplications in $\mathbb{F}_{3^{12}}$ (resp. $\mathbb{F}_{3^6}$) can be performed in 1 s (cf. Sect. 5.1).

Finding logarithms of linear polynomials		
Relation generation	$2^{28.6} M_{q^2}$	$2^{28.6} M_{q^2}$
Linear algebra	$2^{47.5} A_r$	$2^{49.5} M_{q^2}$
Finding logarithms of irreducible quadratic polynomials		
Relation generation	$3^{12} \times 2^{37.6} M_{q^2}$	$2^{56.6} M_{q^2}$
Linear algebra	$3^{12} \times 2^{47.5} A_r$	$2^{68.5} M_{q^2}$
Descent		
Continued-fractions (714 to 88)	$2^{77.6} M_q$	$2^{77.6} M_q$
Classical (88 to 29)	$16.2 \times 2^{73.5} M_q$	$2^{77.5} M_q$
Classical (29 to 11)	$267.3 \times 2^{70.8} M_q$	$2^{78.9} M_q$
QPA (11 to 7)	$2^{13.9} \times (2^{44.4} M_{q^2} + 2^{47.5} A_r)$	$2^{63.4} M_{q^2}$
Gröbner bases (7 to 4)	$2^{35.2} \times (76.9\text{s})$	$2^{67.5} M_{q^2}$
Gröbner bases (4 to 3)	$2^{44.7} \times (0.03135\text{s})$	$2^{65.7} M_{q^2}$
Gröbner bases (3 to 2)	$2^{54.2} \times (0.002532\text{s})$	$2^{71.6} M_{q^2}$

The new cost estimates are presented in Table 2. The main effect of the techniques from [18] is the removal of one QPA descent stage. The overall running time is $2^{78.8} M_{q^2}$, a significant improvement over the $2^{95.8} M_{q^2}$ estimate from [2].

6   Conclusions

We used Joux's algorithm to solve instances of the discrete logarithm problem in the 1303-bit finite field $\mathbb{F}_{3^{6 \cdot 137}}$ and the 1551-bit finite field $\mathbb{F}_{3^{6 \cdot 163}}$. We emphasize that these fields are 'general' in that they do not enjoy any Kummer-like properties. The computations took only 888 CPU hours and 1201 CPU hours, respectively, using modest computer resources despite our implementation being in Magma and far from optimal, unlike the substantial resources (approximately 800,000 CPU hours) that were consumed in [22] for computing a logarithm in the 923-bit field $\mathbb{F}_{3^{6 \cdot 97}}$ with the Joux-Lercier algorithm. We also used newer techniques from [26] and [18] to lower the estimates for computing discrete logarithms in $\mathbb{F}_{3^{6 \cdot 509}}$ and $\mathbb{F}_{3^{6 \cdot 1429}}$ to $2^{58.9} M_q$ and $2^{78.8} M_{q^2}$ (where $q = 3^6$). Our computational results add further weight to the claim that Joux's algorithm and its quasi-polytime successor [4] render bilinear pairings derived from the supersingular elliptic curves $E : y^2 = x^3 - x \pm 1$ over $\mathbb{F}_{3^n}$ unsuitable for pairing-based cryptography.

A   Magma Script for Verifying the $\mathbb{F}_{3^{6 \cdot 137}}$ discrete log

```
//Definition of the extension fields Fq := F3(U) and Fq3 := Fq(V)
q        := 3^4;
F3       := FiniteField(3);
P3<u>    := PolynomialRing(F3);
poly     := u^4 + u^2 + 2;
Fq<U>    := ext<F3|poly>;
Pq<v>    := PolynomialRing(Fq);
poly     := v^3 + v + U^2 + U;
Fq3<V>   := ext<Fq|poly>;
Pq3<Z>   := PolynomialRing(Fq3);
r        := 3309828011909019102877558005508217506428495623;
Fr       := GF(r);

h0       := V^326196*Z^2 + V^35305*Z + V^204091;
h0q      := Evaluate(h0,Z^q);
F        := Z - h0q;
Ix       := Factorization(F)[2][1];
Fn<X>    := ext<Fq3|Ix>;
N        := #Fn - 1;

// Generator of GF(3^{12*137})^*
g        := X + V^113713;
// Encoding pi
Re       := RealField(2000);
pival    :=Pi(Re);
hp       := 0;
```

```
for i := 0 to 136 do
        hp := hp + V^(Floor(pival*(#Fq3)^(i+1)) mod #Fq3)*(X^i);
end for;
// This is the logarithm challenge
cofactor := N div r;
h         := hp^cofactor;


// log_g(h) mod r is:
x         := 2733961907697509392024551597321418696302565659;


// Define the exponent y to be used in the verification:
y         := IntegerRing()!(Fr!(x/cofactor));


// Check that h = (g^cofactor)^y
h eq (g^cofactor)^y;
```

B   Magma Script for Verifying the $\mathbb{F}_{3^{6 \cdot 163}}$ discrete log

```
//Definition of the extension fields Fq := F3(U) and Fq3 := Fq(V)
q         := 3^4;
F3        := FiniteField(3);
P3<u>     := PolynomialRing(F3);
poly      := u^4 + u^2 + 2;
Fq<U>     := ext<F3|poly>;
Pq<v>     := PolynomialRing(Fq);
poly      := v^3 + v + U^2 + U;
Fq3<V>    := ext<Fq|poly>;
Pq3<Z>    := PolynomialRing(Fq3);
r         := 5898811514266587408542277255807363488506406322973734140917909
             955505756623268837;
Fr        := GF(r);


h1        := Z^2 + V^530855;
h1q       := Evaluate(h1,Z^q);
Ix        := h1q*Z - 1;
Fn<X>     := ext<Fq3|Ix>;
N         := #Fn - 1;


// Generator of GF(3^{12*163})^*
g         := X + V^2;


// Encoding pi
Re        := RealField(2000);
pival     :=Pi(Re);
hp        := 0;
for i := 0 to 162 do
        hp := hp + V^(Floor(pival*(#Fq3)^(i+1)) mod #Fq3)*(X^i);
end for;
```

```
// This is the logarithm challenge
cofactor := N div r;
h        := hp^cofactor;

// log_g(h) mod r is:
x        := 4263959514982791937132913919534490007325925542511325256720
            39784356054526194343;

// Define the exponent y to be used in the verification:
y        := IntegerRing()!(Fr!(x/cofactor));

// Check that h = (g^cofactor)^y
h eq (g^cofactor)^y;
```

References

1. Adj, G., Menezes, A., Oliveira, T., Rodríguez-Henríquez, F.: Weakness of $\mathbb{F}_{3^{6\cdot509}}$ for discrete logarithm cryptography. In: Cao, Z., Zhang, F. (eds.) Pairing 2013. LNCS, vol. 8365, pp. 20–44. Springer, Heidelberg (2014)
2. Adj, G., Menezes, A., Oliveira, T., Rodríguez-Henríquez, F.: Weakness of $\mathbb{F}_{3^{6\cdot1429}}$ and $\mathbb{F}_{2^{4\cdot3041}}$ for discrete logarithm cryptography. Finite Fields and Their Applications (to appear)
3. Barbulescu, R., Bouvier, C., Detrey, J., Gaudry, P., Jeljeli, H., Thomé, E., Videau, M., Zimmermann, P.: Discrete logarithm in GF(2^{809}) with FFS. In: Krawczyk, H. (ed.) PKC 2014. LNCS, vol. 8383, pp. 221–238. Springer, Heidelberg (2014)
4. Barbulescu, R., Gaudry, P., Joux, A., Thomé, E.: A heuristic quasi-polynomial algorithm for discrete logarithm in finite fields of small characteristic. In: Nguyen, P.Q., Oswald, E. (eds.) EUROCRYPT 2014. LNCS, vol. 8441, pp. 1–16. Springer, Heidelberg (2014)
5. Barreto, P.S.L.M., Kim, H.Y., Lynn, B., Scott, M.: Efficient algorithms for pairing-based cryptosystems. In: Yung, M. (ed.) CRYPTO 2002. LNCS, vol. 2442, pp. 354–368. Springer, Heidelberg (2002)
6. Beuchat, J., Detrey, J., Estibals, N., Okamoto, E., Rodríguez-Henríquez, F.: Fast architectures for the η_T pairing over small-characteristic supersingular elliptic curves. IEEE Trans. Comput. 60, 266–281 (2011)
7. Blake, I., Fuji-Hara, R., Mullin, R., Vanstone, S.: Computing logarithms in finite fields of characteristic two. SIAM J. Algebraic Discrete Methods 5, 276–285 (1984)
8. Boneh, D., Lynn, B., Shacham, H.: Short signatures from the Weil pairing. J. Cryptology 17, 297–319 (2004)
9. Coppersmith, D.: Fast evaluation of logarithms in fields of characteristic two. IEEE Trans. Inf. Theory 30, 587–594 (1984)
10. Coppersmith, D.: Solving homogeneous linear equations over $GF(2)$ via block Wiedemann algorithm. Math. Comput. 62, 333–350 (1994)
11. The Cunningham Project. http://homes.cerias.purdue.edu/ssw/cun/
12. Faugère, J.: A new efficient algorithm for computing Gröbner bases (F_4). J. Pure Appl. Algebra 139, 61–88 (1999)
13. Frey, G., Rück, H.: A remark concerning m-divisibility and the discrete logarithm in the divisor class group of curves. Math. Comput. 62, 865–874 (1994)

14. Galbraith, S.D., Harrison, K., Soldera, D.: Implementing the tate pairing. In: Fieker, C., Kohel, D.R. (eds.) ANTS 2002. LNCS, vol. 2369, pp. 324–337. Springer, Heidelberg (2002)
15. Göloğlu, F., Granger, R., McGuire, G., Zumbrägel, J.: On the function field sieve and the impact of higher splitting probabilities. In: Canetti, R., Garay, J.A. (eds.) CRYPTO 2013, Part II. LNCS, vol. 8043, pp. 109–128. Springer, Heidelberg (2013)
16. Göloğlu, F., Granger, R., McGuire, G., Zumbrägel, J.: Solving a 6120-bit DLP on a desktop computer. In: Lange, T., Lauter, K., Lisoněk, P. (eds.) SAC 2013. LNCS, vol. 8282, pp. 136–152. Springer, Heidelberg (2014)
17. Granger, R., Kleinjung, T., Zumbrägel, J.: Breaking '128-bit secure' supersingular binary curves (or how to solve discrete logarithms in $\mathbb{F}_{2^{4 \cdot 1223}}$ and $\mathbb{F}_{2^{12 \cdot 367}}$). http://eprint.iacr.org/2014/119
18. Granger, R., Kleinjung, T., Zumbrägel, J.: Breaking '128-bit Secure' supersingular binary curves. In: Garay, J.A., Gennaro, R. (eds.) CRYPTO 2014, Part II. LNCS, vol. 8617, pp. 126–145. Springer, Heidelberg (2014)
19. Granger, R., Page, D., Stam, M.: Hardware and software normal basis arithmetic for pairing based cryptography in characteristic three. IEEE Trans. Comput. **54**, 852–860 (2005)
20. Granger, R., Page, D., Stam, M.: On small characteristic algebraic tori in pairing-based cryptography. LMS J. Comput. Math. **9**, 64–85 (2006)
21. Granger, R., Zumbrägel, J.: On the security of supersingular binary curves. presentation at ECC 2013 (16 September 2013)
22. Hayashi, T., Shimoyama, T., Shinohara, N., Takagi, T.: Breaking pairing-based cryptosystems using η_T pairing over $GF(3^{97})$. In: Wang, X., Sako, K. (eds.) ASIACRYPT 2012. LNCS, vol. 7658, pp. 43–60. Springer, Heidelberg (2012)
23. Joux, A.: A new index calculus algorithm with complexity $L(1/4 + o(1))$ in small characteristic. In: Lange, T., Lauter, K., Lisoněk, P. (eds.) SAC 2013. LNCS, vol. 8282, pp. 355–380. Springer, Heidelberg (2014)
24. Joux, A.: Discrete logarithm in $GF(2^{6128})$, Number Theory List (21 May 21 2013)
25. Joux, A., Lercier, R.: The function field sieve in the medium prime case. In: Vaudenay, S. (ed.) EUROCRYPT 2006. LNCS, vol. 4004, pp. 254–270. Springer, Heidelberg (2006)
26. Joux, A., Pierrot, C.: Improving the polynomial time precomputation of frobenius representation discrete logarithm algorithms. In: Sarkar, P., Iwata, T. (eds.) ASIACRYPT 2014. LNCS, vol. 8873, pp. 378–397. Springer, Heidelberg (2014)
27. Magma v2.19-7. http://magma.maths.usyd.edu.au/magma/
28. Menezes, A., Okamoto, T., Vanstone, S.: Reducing elliptic curve logarithms to logarithms in a finite field. IEEE Trans. Inf. Theory **39**, 1639–1646 (1993)
29. Pollard, J.: Monte Carlo methods for index computation mod p. Math. Comput. **32**, 918–924 (1978)
30. Shinohara, N., Shimoyama, T., Hayashi, T., Takagi, T.: Key length estimation of pairing-based cryptosystems using η_T pairing. In: Ryan, M.D., Smyth, B., Wang, G. (eds.) ISPEC 2012. LNCS, vol. 7232, pp. 228–244. Springer, Heidelberg (2012)
31. Wiedemann, D.: Solving sparse linear equations over finite fields. IEEE Trans. Inf. Theory **32**, 54–62 (1986)

Finite Field Arithmetic

Accelerating Iterative SpMV for the Discrete Logarithm Problem Using GPUs

Hamza Jeljeli[✉]

CARAMEL project-team, LORIA, INRIA/CNRS/Université de Lorraine,
Campus Scientifique, BP 239, 54506 Vandœuvre-lés-Nancy Cedex, France
Hamza.Jeljeli@loria.fr

Abstract. In the context of cryptanalysis, computing discrete logarithms in large cyclic groups using index-calculus-based methods, such as the number field sieve or the function field sieve, requires solving large sparse systems of linear equations modulo the group order. Most of the fast algorithms used to solve such systems — e.g., the conjugate gradient or the Lanczos and Wiedemann algorithms — iterate a product of the corresponding sparse matrix with a vector (SpMV). This central operation can be accelerated on GPUs using specific computing models and addressing patterns, which increase the arithmetic intensity while reducing irregular memory accesses. In this work, we investigate the implementation of SpMV kernels on NVIDIA GPUs, for several representations of the sparse matrix in memory. We explore the use of Residue Number System (RNS) arithmetic to accelerate modular operations. We target linear systems arising when attacking the discrete logarithm problem on groups of size 100 to 1000 bits, which includes the relevant range for current cryptanalytic computations. The proposed SpMV implementation contributed to solving the discrete logarithm problem in $GF(2^{619})$ and $GF(2^{809})$ using the FFS algorithm.

Keywords: Discrete logarithm problem · Sparse-matrix–vector product · Modular arithmetic · Residue number system · GPUs

1 Introduction

The security of many cryptographic protocols used for authentication, key exchange, encryption, or signature, depends on the difficulty of solving the discrete logarithm problem (DLP) in a given cyclic group [16]. For instance, we can rely on the hardness of the DLP in a multiplicative subgroup of a finite field. There are algorithms, such as Pollard-rho [17] or Baby-Step/Giant-Step [21] that solve the problem in time exponential in the subgroup size. Another family of methods, known as *Index-calculus* methods [1] propose to solve it in time subexponential or quasi-polynomial in the finite field size. These algorithms require in their linear algebra step the resolution of large sparse systems of linear equations modulo the group order [12]. In cryptographic applications, the group order ℓ is of size 100 to 1000 bits. The number of rows and columns of the corresponding

© Springer International Publishing Switzerland 2015
Ç. Koç et al. (Eds.): WAIFI 2014, LNCS 9061, pp. 25–44, 2015.
DOI: 10.1007/978-3-319-16277-5_2

matrices is in the order of hundreds of thousands to millions, with only hundreds or fewer non-zero elements per row. This linear algebra step is a serious limiting factor in such algorithms. For example, it was reported in [9] that the linear algebra step of the Function Field Sieve (FFS) implementation to solve the DLP over $GF(3^{6 \times 97})$ took 80.1 days on 252 CPU cores, which represents 54 % of the total time.

To solve such systems, ordinary Gaussian elimination is inefficient. While some elimination strategies aiming at keeping the matrix as sparse as possible can be used to reduce the input system somewhat, actual solving calls for the use of other techniques (Lanczos algorithm [13], Wiedemann algorithm [27]) that take advantage of the sparsity of the matrix [18]. For the Lanczos algorithm, the Wiedemann algorithm and their block variants, the iterative sparse-matrix–vector product is the most time-consuming operation. For this reason, we investigate accelerating this operation on GPUs.

The paper is organized as follows. Section 2 presents the background related to the hardware and the context. Section 3 discusses the arithmetic aspects of our implementation. We present several matrix formats and their corresponding implementations in Sect. 4. We discuss in Sect. 5 how to adapt these implementations over large fields. We compare the results of different implementations run on NVIDIA GPUs in Sect. 6, and present optimizations based on hardware considerations in Sect. 7. Section 8 discusses our reference software implementation.

2   Background

2.1   GPUs and the CUDA Programming Model

CUDA is the hardware and software architecture that enables NVIDIA GPUs to execute programs written in C, C++, OpenCL and other languages [14].

A CUDA program instantiates a *host* code running on the CPU and a *kernel* code running on the GPU. The kernel code runs according to the Single Program Multiple Threads (SPMT) execution model across a set of parallel threads. The threads are executed in groups of 32, called *warps*. If one or more threads have a different execution path, execution divergence occurs. The different paths will then be serialized, negatively impacting the performance.

The threads are further organized into thread *blocks* and *grids* of thread blocks:

- A thread executes an instance of the kernel. It has a unique thread ID within its thread block, along with registers and private memory.
- A thread block is a set of threads executing the same kernel which can share data through *shared memory* and perform barrier synchronization which ensures that all threads within that block reach the same instruction before continuing. It has a unique block ID within its grid.
- A grid is an array of thread blocks executing the same kernel. All the threads of the grid can also read inputs, and write results to *global memory*.

At the hardware level, the blocks are distributed on an array of multi-core *Streaming Multiprocessors* (SMs). Each SM schedules and launches the threads in groups of warps. Recent NVIDIA GPUs of family name *"Kepler"* allow for up to 64 active warps per SM. The ratio of active warps to the maximum supported is called *occupancy*. Maximizing the occupancy is important, as it helps to hide the memory latency. One should therefore pay attention to the usage of shared memory and registers in order to maximize occupancy.

Another important performance consideration in programming for the CUDA architecture is *coalescing* global memory accesses. To understand this requirement, global memory should be viewed in terms of aligned segments of 32 words of 32 bits each. Memory requests are serviced for one warp at a time. If the warp requests hit exactly one segment, the access is *fully coalesced* and there will be only one memory transaction performed. If the warp accesses scattered locations, the accesses are *uncoalesced* and there will be as many transactions as the number of hit segments. Consequently, a kernel should use a coalescing-friendly pattern for greater memory efficiency.

Despite their high arithmetic intensity and their large memory bandwidth, GPUs provide small caches. In fact, Kepler GPUs provide the following levels of cache:

- 1536-kB *L2-cache* per GPU.
- 16-kB *L1-cache* (per SM). It can be extended to 48 kB, but this decreases shared memory from 48 kB to 16 kB.
- A *texture cache*: an on-chip cache for the read-only *texture memory*. It can accelerate memory accesses when neighboring threads read from nearby addresses.

2.2 Sparse-Matrix–Vector Product on GPUs

Sparse-matrix computations pose some difficulties on GPUs, such as irregular memory accesses, load balancing and low cache efficiency. Several papers have focused on choosing suitable matrix formats and appropriate kernels to overcome the irregularity of the sparse matrix [4,26]. These works have explored implementing efficiently SpMV over real numbers. Schmidt et al. [19] proposed an optimized matrix format to accelerate exact SpMV over GF(2), that can be used in the linear algebra step of the Number Field Sieve (NFS) for integer factorization [22]. Boyer et al. [8] have adapted SpMV kernels over small finite fields and rings $\mathbb{Z}/m\mathbb{Z}$, where they used double-precision floating-point numbers to represent ring elements. In our context, since the order of the considered finite ring is large (hundreds of bits), specific computing models and addressing models should be used.

In this work, we have a prime ℓ, along with an N-by-N sparse matrix A defined over $\mathbb{Z}$, and we want to solve the linear system $Aw = 0$ over $\mathbb{Z}/\ell\mathbb{Z}$. A feature of the index calculus context that we consider here, is that A contains small values (e.g. 32-bit integers). In fact, around 90 % of the non-zero coefficients are ± 1.

The very first columns of A are relatively dense, then the column density decreases gradually. The row density does not change significantly. We denote by n_{NZ} the number of non-zero elements in A. See Fig. 1 for a typical density plot of a matrix arising in an FFS computation.

We will use the Wiedemann algorithm as a solver. This algorithm iterates a very large number of matrix-vector products of the form $v \leftarrow Au$, where u and v are dense N-coordinate vectors. The major part of this work deals with how to accelerate this product.

Fig. 1. Distribution of non-zero elements in an FFS matrix

In order to carry out this product, we compute the dot product between each row of A and the vector u. The basic operation is of the form $x \leftarrow (x + \lambda y) \bmod \ell$, where λ is a non-zero coefficient of A, and x and y are coordinates of the vectors v and u, respectively. To minimize the number of costly reductions modulo ℓ, we accumulate computations, and postpone the final modular reduction of the result as late as possible. When iterating many products (computations of the form $A^i u$), we can further accumulate several SpMVs before reducing modulo ℓ, as long as the intermediate results do not exceed the largest representable integer. As far as arithmetic over $\mathbb{Z}/\ell\mathbb{Z}$ is concerned, we chose to use the Residue Number System, which appears to be more suited to the fine grained parallelism inherent to the SPMT computing model than the usual multi-precision representation of large integers. A comparison of the two representations is given in Subsect. 6.3.

3   Residue Number System and Modular Arithmetic

3.1   A Brief Reminder on RNS

The Residue Number System (RNS) is based on the Chinese Remainder Theorem (CRT). Let $\mathcal{B} = (p_1, p_2, \ldots, p_n)$ be a set of mutually coprime integers, which we call an *RNS-basis*. We define P as the product of all the p_i's. The RNS uses the fact that any integer x within $[0, P-1]$ can be uniquely represented by the list $(x_1, x_2, \ldots, x_n)$, where each x_i is the residue of x modulo p_i, which we write as $x_i = |x|_{p_i}$.

If x and y are given in their RNS representations $\vec{x} = (x_1, \ldots, x_n)$ and $\vec{y} = (y_1, \ldots, y_n)$, according to $\mathcal{B}$, and such that $x, y < P$, RNS addition and multiplication are realized by modular addition and multiplication on each component:

$$\vec{x} \vec{+} \vec{y} = (|x_1 + y_1|_{p_1}, \ldots, |x_n + y_n|_{p_n}), \quad \vec{x} \vec{\times} \vec{y} = (|x_1 \times y_1|_{p_1}, \ldots, |x_n \times y_n|_{p_n})$$

The result (e.g., $x+y$) should belong to the interval $[0, P-1]$ if we want to obtain a valid RNS representation. Otherwise, it will be reduced modulo P. Unlike addition or multiplication, other operations such as comparison or modular reduction are more subtle in RNS.

We can convert back an RNS vector to the integer form by using the CRT formula:

$$x = \left| \sum_{i=1}^{n} x_i \cdot \left| P_i^{-1} \right|_{p_i} \cdot P_i \right|_P \quad , \text{where } P_i \triangleq P/p_i.$$

This number system is particularly interesting for arithmetic over large integers, since it distributes the computation over several small residues. In other words, the computation units that will work on the residues are independent and need no synchronization nor communication, as there is no carry propagation [23, 24].

3.2   RNS Reduction Modulo ℓ

In the chosen RNS representation, $(P - 1)$ is the largest representable integer. So in the case of repeated SpMVs over $\mathbb{Z}/\ell\mathbb{Z}$, we can accumulate at most $\log(\frac{P-1}{\ell-1})/\log(r)$ matrix–vector products before having to reduce modulo ℓ, where r corresponds to the largest row norm (defined as the sum of the absolute values of its elements) in the matrix. To reduce the vector v modulo ℓ, we use the method introduced by Bernstein in [6], which allows us to perform the reduction without having to convert the vector back to the integer form.

We assume that the RNS-basis $\mathcal{B}$ contains n moduli $p_1, \ldots, p_n$ of k bits each. We impose that the p_i's are close to 2^k. The reasons will be detailed in the following subsection. We want to reduce modulo ℓ an RNS vector $(x_1, \ldots, x_n)$.

We start from the CRT reconstruction: $x = \left| \sum_{i=1}^{n} \gamma_i P_i \right|_P$, where we have defined $\gamma_i \triangleq \left| x_i P_i^{-1} \right|_{p_i}$. Let us also define the integer α as follows

$$\alpha = \left\lfloor \sum_{i=1}^{n} \frac{\gamma_i P_i}{P} \right\rfloor = \left\lfloor \sum_{i=1}^{n} \frac{\gamma_i}{p_i} \right\rfloor. \tag{1}$$

The vector x can then be written as $\sum_{i=1}^{n} \gamma_i P_i - \alpha P$ and, since $\gamma_i < p_i$, we have $0 \leq \alpha < n$.

Now, if we assume that α is known, we define $z \triangleq \sum_{i=1}^{n} \gamma_i \left| P_i \right|_\ell - \left| \alpha P \right|_\ell$. We can easily check that z is congruent to x modulo ℓ and lies in the interval $[0, \ell \sum_{i=1}^{n} p_i[$.

What remains to be done is to determine α. Since $p_i \approx 2^k$, we approximate the quotient γ_i/p_i using only the s most significant bits of $\gamma_i/2^k$. Hence, an estimate for α is proposed as

$$\hat{\alpha} \triangleq \left\lfloor \sum_{i=1}^{n} \frac{\left\lfloor \frac{\gamma_i}{2^{k-s}} \right\rfloor}{2^s} + \Delta \right\rfloor, \tag{2}$$

where s is an integer parameter in $[1, k]$ and Δ an error correcting term in $]0, 1[$.

Bernstein states in [6] that if $0 \leq x < (1 - \Delta)P$ and $(\epsilon + \delta) \leq \Delta < 1$ where

$$\epsilon \triangleq \sum_{i=1}^{n} \frac{c_i}{2^k} \text{ and } \delta \triangleq n \frac{2^{k-s} - 1}{2^k}, \text{ then } \alpha = \hat{\alpha}.$$

Once α is determined, we are able to perform an RNS computation of z. Algorithm 1 summarizes the steps of the computation.

Algorithm 1. Approximate RNS modular reduction

Precomputed: Vector $\left(\left|P_j^{-1}\right|_{p_j}\right)$ for $j \in \{1, \ldots, n\}$

Table of RNS vectors of $|P_i|_\ell$ for $i \in \{1, \ldots, n\}$

Table of RNS vectors of $|\alpha P|_\ell$ for $\alpha \in \{1, \ldots, n-1\}$

Input   : RNS vector of x, with $0 \leq x < (1 - \Delta)P$

Output   : RNS vector of $z \equiv x \pmod{\ell}$, $z < \ell \sum_{i=1}^{n} p_i$

1 **foreach** *thread j* **do**

2 $\quad \gamma_j \leftarrow \left|x_j \times \left|P_j^{-1}\right|_{p_j}\right|_{p_j}$ /* 1 RNS product */

3 Broadcast of the γ_j's by all the threads

4 **foreach** *thread j* **do**

5 $\quad z_j \leftarrow \left|\sum_{i=1}^{n} \gamma_i \times ||P_i|_\ell|_{p_j}\right|_{p_j}$ /* $(n-1)$ RNS sums & n RNS products */

6 $\quad \alpha \leftarrow \left\lfloor \sum_{i=1}^{n} \frac{\left\lfloor \frac{\gamma_i}{2^{k-s}} \right\rfloor}{2^s} + \Delta \right\rfloor$ /* sum of n s-bit terms */

7 $\quad z_j \leftarrow \left|z_j - ||\alpha P|_\ell|_{p_j}\right|_{p_j}$ /* 1 RNS subtraction */

All the operations can be evaluated in parallel on the residues, except for step 3, where a broadcast of all the γ_j's is needed. Even if the obtained result z is not the exact reduction of x, it is bounded by $n 2^k \ell$. Thus, we guarantee that the intermediate results of the SpMV computation do not exceed a certain bound less than P. Notice that this RNS reduction algorithm imposes that P be one modulus (k bits) larger than implied by the earlier condition $\ell < P$.

In conclusion, P is chosen, such that $r \times n 2^k \ell < (1 - \Delta)P$, with r is the largest row norm of the matrix.

3.3 Modular Reduction Modulo p_j

The basic RNS operation is $z_j \leftarrow (x_j + \lambda \times y_j) \bmod p_j$, where $0 \leq x_j, y_j, z_j < p_j$ are RNS residues and λ is a positive element of the matrix. So, it consists of an AddMul (multiplication, then an addition) followed by a reduction modulo p_j.

To speed up the reduction modulo p_j, the moduli are chosen of the pseudo-Mersenne form $2^k - c_j$, with c_j as small as possible.

In fact, let us define $t_j \triangleq x_j + \lambda \times y_j$ as the intermediate result before the modular reduction. t_j can be written as

$$t_j = t_{jL} + 2^k \times t_{jH}, \text{ where } t_{jL} \triangleq t_j \bmod 2^k, t_{jH} \triangleq t_j/2^k. \qquad (3)$$

Since $2^k \equiv c_j \pmod{p_j}$, we have $t_j \equiv t_{jL} + t_{jH} \times c_j \pmod{p_j}$. So, we compute $t_j \leftarrow t_{jL} + t_{jH} \times c_j$, then we have to consider two cases:

- if $t_j < 2^k$, we have "almost" reduced $(x_j + \lambda \times y_j)$ modulo p_j, since the result lies in $[0, 2^k[$, not in $[0, p_j[$;
- else we have reduced t_j by approximately k bits. Thus, we repeat the previous procedure with $t_j \leftarrow t_{jL} + c_j \times t_{jH}$, which then satisfies $t_j < 2^k$.

The output lies in $[0, 2^k - 1]$, so we propose to relax the condition on both input and output: $x_j, z_j \in [0, 2^k - 1]$. With this approach, the reduction can be done in a small number of additions and products.

3.4 Possible RNS Mappings on GPU/CPU

We represent the finite ring $\mathbb{Z}/\ell\mathbb{Z}$ as the integer interval $[0, \ell - 1]$. Each element is stored in its RNS form. On GPU, we opted for 64-bit moduli (i.e. $k = 64$), for performance considerations. Even that floating point instructions have higher throughput, integer instructions gave better performances, because with floating point arithmetic, only the mantissa is used and the algorithms are more complex than with integer arithmetic. We use the PTX (*parallel thread execution*) pseudo-assembly language for CUDA [15] to implement the RNS operations.

On CPU, we implemented three versions based on:

- MMX instruction set: we map an RNS residue to an unsigned 64-bit integer.
- Streaming SIMD Extensions (SSE2) set: a 128-bit XMM register holds two residues, so the processor can process two residues simultaneously.
- Advanced Vector Extensions (AVX2) set: we use the 256-bit YMM register to hold four residues.

4 Sparse Matrix Storage Formats

In this section, we assume that the elements of the matrix, as well as the elements of the vectors u and v are in a field K (reals, finite fields, etc.). For each format, we will discuss how to perform the matrix-vector product. We will give a pseudo-code for the format CSR. Figures that illustrate the other formats and their corresponding Pseudo-code can be found in Appendix A.

The matrix and vectors are put in *global memory*, since their sizes are important. Temporary results are stored in registers. The *shared memory* is used when partial results of different threads are combined. Arithmetic operations are performed in registers and denoted in the pseudo-code by the function addmul().

Coordinate (COO). The format COO consists of three arrays `row_id`, `col_id` and `data` of n_{NZ} elements. The row index, column index and the value are explicitly stored to specify a non-zero matrix coefficient. In this work, we propose to sort the matrix coefficients by their row index.

A typical way to work with the COO format on GPU is to assign one thread to each non-zero matrix coefficient. This implies that different threads from different warps will process a same row. Each thread computes its partial result, then performs a segmented reduction [7,20] to sum the partial results of the other threads belonging to the same warp and spanning the same row. We followed the scheme proposed by the library CUSP [5], which performs the segmented reduction in shared memory, using the row indices as segment descriptors. Each warp iterates over its interval, processing 32 coefficients at a time. If a spanned row is fully processed, its result is written to v, otherwise, the row index and the partial dot product are stored in temporary arrays. Then, a second kernel performs the combination of the per-warp results.

The main drawbacks of the COO kernel are the cost of the combination of partial results and excessive usage of *global memory*. Its advantage is that the workload distribution is balanced across warps, as they iterate over a constant length interval.

Compressed Sparse Row (CSR). The CSR format stores the column indices and the values of the non-zero elements of A into two arrays of n_{NZ} elements: `id` and `data`. A third array of pointers, `ptr`, of length $N + 1$, is used to indicate the beginning and the end of each row. Non-zero coefficients are sorted by their row index. The CSR format eliminates the explicit storage of the row index, and is convenient for a direct access to the matrix, since `ptr` indicates where each row starts and ends in the other two ordered arrays.

$$
\begin{pmatrix}
0 & a_{01} & 0 & a_{03} & 0 & 0 \\
0 & a_{11} & 0 & 0 & a_{14} & a_{15} \\
a_{20} & 0 & a_{22} & a_{23} & 0 & 0 \\
0 & a_{31} & 0 & 0 & a_{34} & 0 \\
0 & a_{41} & a_{42} & 0 & 0 & a_{45} \\
0 & 0 & a_{52} & 0 & 0 & a_{55}
\end{pmatrix}
$$

(a) Sparse matrix A

$$
\text{data} - \begin{bmatrix} a_{01} & a_{03} & a_{11} & a_{14} & a_{15} & a_{20} & a_{22} & a_{23} & \cdots \end{bmatrix}
$$

$$
\text{id} - \begin{bmatrix} 1 & 3 & 1 & 4 & 5 & 0 & 2 & 3 & \cdots \end{bmatrix}
$$

$$
\text{ptr} - \begin{bmatrix} 0 & 2 & 5 & 8 & \cdots \end{bmatrix}
$$

(b) CSR representation

Scalar Approach (CSR-S). To parallelize the product for the CSR format, a simple way is to assign each row to a single thread (*scalar* approach). For each non-zero coefficient, the thread performs a read from *global memory*, an `addmul` and a write in *registers*. Final result is written to *global memory*.

Vector Approach (CSR-V). The *vector* approach consists in assigning a warp to each row of the matrix [4]. The threads within a warp access neighboring non-zero elements, which makes the warp accesses to id and data contiguous. Each thread computes its partial result in shared memory, then a parallel reduction in shared memory is required to combine the per-thread results (denoted reduction_csr_v() in Algorithm 2). No synchronization is needed, since threads belonging to a same warp are implicitly synchronized.

Algorithm 2. CSR-V for row i executed by thread of index tid in its warp

---

Inputs : data: array of n_{NZ} elements of K, id: array of n_{NZ} positive integers,
        ptr: array of N positive integers and u: vector of N elements of K.
Output: v: vector of N elements of K.

sum $\leftarrow 0$;
$j \leftarrow$ ptr$_i$ + tid;   // position of beginning for each thread
While $j <$ ptr$_{i+1}$ do
  | sum $\leftarrow$ addmul(sum,data$_j$,u_{id_j});
  | $j \leftarrow j + 32$;
reduction_csr_v(sum,tid);   // reduction in *shared memory*
If tid= 0 then   // first thread of the warp writes in *global memory*
  | $v_i \leftarrow$ sum;

Compared to COO kernel, the two CSR kernels reduce the usage of *global memory* and simplify the combination of partial results. The CSR kernels suffer from load unbalance, if the rows have widely varying lengths. To improve the load balance, one possibility is to order the rows by their lengths. So, the warps launched simultaneously have almost the same load.

If we compare the two CSR kernels. The threads of CSR-S have non contiguous access to data et id, as they do not work on the same rows. Thus, their memory accesses are not as efficient as the accesses of the CSR-V. However, the CSR-V kernel requires a combination of partial results which increases the use of *registers* and *shared memory* (cf. Subsect. 6.2).

ELLPACK (ELL). The ELL format extends the CSR arrays to N-by-K arrays, where K corresponds to the maximum number of non-zero coefficients per row. The rows that have less than K non-zero coefficients are padded. Since the padded rows have the same length, only column indices are explicitly stored. This format suffers from the overhead due to the padding when the percentage of zeros is high. An optimization was proposed by Vázquez et al. with a format called ELLPACK-R (ELL-R) [26]. This variant adds an array len of length N that indicates the length of each row. Thus, the zeros added by the padding are not considered when performing the matrix-vector product.

The partitioning of the work is done by assigning a thread to a row of the matrix. The kernel takes advantage from the column-major ordering of the elements to improve the accesses on the vector u. However, it suffers from thread divergence.

Sliced Coordinate (SLCOO). The SLCOO format was introduced on GPUs by Schmidt et al. for integer factorization, in the particular case of matrices over GF(2) [19] and was inspired by the CADO-NFS [2] software for CPUs. The aim of this format is to increase the cache hit rate that limits the CSR and COO performance. Like COO, the SLCOO representation stores the row indices, column indices and values. However, it divides the matrix into horizontal slices, where the non-zero coefficients of a slice are sorted according to their column index in order to reduce the irregular accesses on source vector u, if they had been sorted by their row indices. A fourth array `ptrSlice` indicates the beginning and end of each slice. We denote this format SLCOO-σ, where the parameter σ is the number of rows in a slice.

For the SLCOO kernel, each warp works on a slice. Since each thread works on more than one row, it needs to have individual storage for its partial per-row results, or to be able to have exclusive access to a common resource. In [19], Schmidt et al. mentionned the two possibilities of either using the *shared memory* or having atomic accesses. While these needs can be fulfilled in [19] for the context of linear algebra over GF(2), we will observe in Sect. 6 that these constraints hamper the efficiency of the SLCOO in the context of large fields.

There are other SpMV formats in the literature, such as DIA (Diagonal) format, that are appropriate only for matrices that satisfy some sparsity patterns, which is not our case.

5   SpMV Kernels Over Large Fields

In the context of our application, the matrix elements are "small" (32 bit integers) and the vectors elements are in $\mathbb{Z}/\ell\mathbb{Z}$. In this section, we study how we adapt the kernels to this context. We assume that an element of $\mathbb{Z}/\ell\mathbb{Z}$ holds in n machine words. Thus, processing a non-zero coefficient λ at row i and column j of the matrix implies reading the n words that compose the jth element in the input vector u, multiply them by λ and adding them to the n words that compose the ith element in the output vector v. In the following pseudo-code, we denote the arithmetic operation that applies to a word by the function `addmul_word()`. Pseudo-code is therefore given without details regarding the representation system of the numbers and the resulting arithmetic.

***Sequential* Scheme.** A first approach would be that each thread processes a coefficient. We would call this scheme *sequential*. To illustrate this scheme, we apply it on the CSR-V kernel.

Algorithm 3. CSR-V-seq for row i executed by thread of index `tid` in its warp

---

Inputs : data: array of n_{NZ} signed integers, id: array of n_{NZ} positive integers,
          ptr: array of N positive integers, u: vector of $N \times n$ machine words.
Output: v: vector of $N \times n$ machine words.

Declare array `sum` $\leftarrow \{0\}$;   // `n machine words initialized to 0`
$j \leftarrow \text{ptr}_i + \text{tid}$;
While $j < \text{ptr}_{i+1}$ **do**
$\quad$ **For** $k \leftarrow 0$ to n **do**
$\quad\quad$ $\text{sum}_k \leftarrow \text{addmul\_word}(\text{sum}_k, \text{data}_j, u_{\text{id}_j \times n + k})$;   // `process` k^{th} `word`
$\quad$ $j \leftarrow j + 32$;
`reduction_csr_v_seq(sum,tid)`;
If `tid`$= 0$ **then**
$\quad$ **For** $k \leftarrow 0$ to n **do**
$\quad\quad$ $v_{i \times n + k} \leftarrow \text{sum}_k$;   // `store` k^{th} `word in` *global memory*

This scheme suffers from several drawbacks. The first one is that the thread processes the n machine words corresponding to the coefficient, i.e. reads and writes the n machine words and makes n arithmetic operations. Thus, the thread consumes more *registers*. The second is that the threads of the same warp access non-contiguous zones of the vectors u and v, as their accesses are always spaced by n words.

Parallel **Scheme.** To overcome the limitations of the previous approach, a better scheme would be that a nonzero coefficient is processed by n threads. We refer to the scheme by the *parallel* scheme. Threads of the same warp are organized in n_{GPS} of n threads, where $n_{\text{GPS}} \times n$ is closest to 32, the number of threads per warp. Each group is associated to a non-zero matrix coefficient. For example, for n = 5, we take $n_{\text{GPS}} = 6$, so the first 5 threads process in parallel the 5 words of the 1st source vector element, threads 5 to 9, process the words of the 2nd source vector element, and so on, and we will have two idle threads per warp.

Algorithm 4. CSR-V-par for row i executed by thread of index `tid` in its warp

---

Inputs : data: array of n_{NZ} signed integers, id: array of n_{NZ} positive integers,
          ptr: array of N positive integers, u: vector of $N \times n$ mots machines.
Output: v: vector of $N \times n$ mots machines.

`sum` $\leftarrow 0$;   // `1 machine word initialized to 0`
$j \leftarrow \text{ptr}_i + \lfloor \text{tid} / n \rfloor$;   // `position of beginning for each thread`
While $j < \text{ptr}_{i+1}$ **do**
$\quad$ `sum` $\leftarrow \text{addmul\_word}(\text{sum}, \text{data}_j, u_{\text{id}_j \times n + \text{tid} \bmod n})$;   // `process 1 word`
$\quad$ $j \leftarrow j + n_{\text{GPS}}$;
`reduction_csr_v_par(sum,tid)`;
If `tid`$< n$ **then**   // `first group of the warp writes in` *global memory*
$\quad$ $v_{i \times n + \text{tid}} \leftarrow \text{sum}$;

For the other kernels, both schemes are applicable and the *parallel* scheme always performs significantly better than the *sequential* scheme.

6 Comparative Analysis of SpMV Kernels

In this section, we compare the performances of the kernels that we presented. The objective is to minimize the time of a matrix-vector product. The experiments were run on an NVIDIA GeForce GTX 680 graphics processor (Kepler). Each SpMV kernel was executed 100 times, which is large enough to obtain stable timings. Our measurements do not include the time spent to copy data between the host and the GPU, since the matrix and vectors do not need to be transferred back and forth between each SpMV iteration. The reduction modulo ℓ happens only once every few iterations, which is why the timing of an iteration includes the timing of the reduction modulo ℓ kernel multiplied by the frequency of its invocation. The reported measurements are based on NVIDIA developer tools.

Table 1 summarizes the considered matrix over $\mathbb{Z}/\ell\mathbb{Z}$. The matrix was obtained during the resolution of discrete logarithm problem in the 217-bit prime order subgroup of $GF(2^{619})^{\times}$ using the FFS algorithm. The $\mathbb{Z}/\ell\mathbb{Z}$ elements fit in four RNS 64-bit residues. Since, an extra residue is needed for the modular reduction (cf. Subsect. 3.2), the total number of RNS residues is $n = 5$.

Table 1. Properties of test matrix

Size of the matrix (N)	650k × 650k
#Non-zero coefficients	65M
Max (row norm)	492
Percentage of ±1	92.7%
Size of ℓ (in bits)	217
Size of M (in bits)	320
Size of $n2^k\ell$ (in bits)	283
Frequency of reduction modℓ	1/4

6.1 Comparison of Schemes *Sequential* and *Parallel*

We compare the application of the two schemes on the CSR-V kernel. The *sequential* kernel consumes more *registers* and *shared memory*, which limits the maximum number of warps that can be run on a SM to 24. In our application, CSR-S was limited to 24 warps/SM, for a bound of 64 warps/SM. This is reported in the column *Theoretical Occupancy* of the following table. The low occupancy significantly decreases the performance. Concerning the *global memory* access pattern, the column *Load/Store efficiency* gives the ratio of requested memory transactions to the number of transactions performed, which reflects the degree to which memory accesses are coalesced (100 % efficiency) or not. For *sequential* kernel, uncoalesced accesses cause the bandwidth loss and the performance degradation. The *parallel* kernel makes the write accesses coalesced (100 % store efficiency). For the loads, it reaches only 47 % due to irregular accesses on the source vector.

	Registers per thread	Shared memory per SM	(Theoretical) occupancy	Load/Store efficiency	Timing in ms
Sequential	27	49152	35.1 % (37.5 %)	7.5 %/26 %	141.1
Parallel	**21**	**15360**	**70.3 % (100 %)**	**47.2 %/100 %**	**41.4**

It is clear that the *parallel* scheme is better suited to the context of large integers. We apply this scheme to other formats and compare their performance in the following subsection.

6.2 Comparison of Kernels CSR, COO, ELL and SLCOO

Due to the segmented reduction, the COO kernel performs more instructions and requires more registers. Thread divergence happens more often, because of the several branches that threads belonging to the same warp can take.

As far as the ELL kernel is concerned, the padded rows have the same length. This yields a good balancing across the warps and the threads (cf. *Occupancy* and *Branch divergence* in the following table). The column-major ordering makes this kernel reach the highest cache hit rate.

The CSR-S kernel suffers from low efficiency of memory access compared to CSR-V. In fact, with the kernel CSR-S, the threads with the same warp work on several lines simultaneously, which makes their access to tables id and data not contiguous. The kernel CSR-V better satisfies the GPU architectural specificities.

	Registers per thread	Branch divergence	(Theoretical) occupancy	Load/Store efficiency	Cache hit rate	Timing in ms
COO	25	47.1 %	(66.7 %) 65.2 %	34.3 %/37.8 %	34.4 %	88.9
CSR-S	18	28.1 %	(100 %) 71.8 %	29.2 %/42.5 %	35.4 %	72.3
CSR-V	**21**	**36.7 %**	**(100 %) 70.3 %**	**47.2 %/100 %**	**35.4 %**	**41.4**
ELL-R	18	44.1 %	(100 %) 71.8 %	38.1 %/42.5 %	40.1 %	45.5

The next table compares the CSR-V kernel with several SLCOO kernels for different slice sizes. We remark that increasing the slice size improves the cache hit rate, since accesses on the source vector are less irregular. However, by making the slices larger, we increase the usage of shared memory proportionally to the slice size, which limits the maximum number of blocks that can run concurrently. This limitation of the occupancy yields poor performance compared to the CSR-V kernel. Here, we consider an L1-oriented configuration (48 kB L1, 16 kB shared) of the on-chip memory. It is possible to move to a shared-oriented configuration (16 kB L1, 48 kB shared). This improves the occupancy, but degrades the cache hit rate, and finally does not improve the performances.

	Shared memory per Block	Blocks per SM	(Theoretical) occupancy	Cache hit rate	Timing in ms
CSR-V	**1920**	**8**	**(100 %) 70.3 %**	**35.4 %**	**41.4**
SLCOO-2	3840	6	(75 %) 68.4 %	36.1 %	46.9
SLCOO-4	7680	4	(50 %) 44.5 %	36.9 %	58.6
SLCOO-8	15360	2	(22.5 %) 22.3 %	37.8 %	89.9

Combinations of different formats have been tested. However they do not give better results. Splitting the matrix into column major blocks, or processing separately the first dense columns did not improve the performance either.

For our matrix, the main bottleneck is memory access. In the CSR-V kernel, 72 % of the time is spent in reading data, 2 % in writing data and 26 % in computations.

6.3   Comparison of RNS and Multi-precision Arithmetics

We also implemented the RNS and the multi-precision (MP) arithmetics on GPU. For the MP representation, to perform the reduction modulo ℓ, we use a precomputed inverse of ℓ so as to divide by ℓ using a single multiplication. For the 217-bit prime order subgroup, choosing the largest representable integer $M = 2^{256} - 1$ is sufficient to accumulate a few number of SpMVs before reducing modulo ℓ. In fact, the maximum row norm that we have (492) allows to do up to 4 iterative SpMVs before having to reduce.

For MP kernel, the reduction kernel takes only 0.37 ms, which corresponds to less than 0.1 ms per iteration. In RNS, we can accumulate 4 SpMVs before the reduction modulo ℓ (cf. Table 1). The reduction kernel takes 1.6 ms (i.e., 0.4 ms per iteration).

The idea behind the use of RNS rather than MP arithmetic is that RNS can significantly decrease data sharing between the threads and arithmetic operations required for the carry generation/propagation. The RNS kernel allows us to reach higher occupancy and better performance. The speed-up of RNS compared to multi-precision on the SpMV timing is around 15 %.

	Registers per thread	Shared memory per Block	Executed instructions	(Theoretical) occupancy	Timing in ms
MP	21	2880	6.1×10^8	(83.3 %) 51.2 %	46.6
RNS	**18**	**1920**	$\mathbf{5.8 \times 10^8}$	**(100 %) 70.3 %**	**41.4**

7   Improvements on CSR-V Kernel

To further improve the kernel performance, one should take into account the GPU architectural characteristics: the management of the memory accesses, the partitioning of the computations and the specificities of the problem considered.

Texture caching. Although our SpMV kernel suffers from irregular load accesses, a thread is likely to read from an address near the addresses that nearby threads (of the same group) read. For this reason, we bind on texture memory and replace reads with texture fetches. This improves the global memory efficiency and consequently the SpMV delay.

Reordering the non-zero coefficients of a row. Since most of the coefficients of the matrix are ± 1, it seems promising to treat multiplications by these coefficients differently from other coefficients: additions and subtractions are less expensive than multiplications. All these separations result in code divergence, that we fix by reordering the non-zero coefficients in the matrix such that values of the same category $(+1, -1, > 0, < 0)$ are contiguous. This decreases the branch divergence and decreases the total SpMV delay.

Compressing the values array `data`. Since the majority of the coefficients are ± 1, after reordering the coefficients per row, we can replace the ± 1 coefficients by their occurrence count. This reduces the length of the values array `data` by more than 10 times, and so reduces the number of reads.

Improving warp balancing. In the CSR-V kernel, each warp processes a single row. This requires launching a large number of warps. Consequently, there is a delay to schedule those launched warps. Instead, we propose that each warp iterates over a certain number of rows. To further increase the occupancy, we permute the rows such that each warp roughly gets the same work load.

	Performance effects	Timing in ms (speedup)
Texture caching	Global Load efficiency: $47.2\% \rightarrow 84\%$	32 (+30%)
Non-zeros reordering	Branch Divergence: $36.7\% \rightarrow 12.9\%$	30.5 (+5%)
Compressing `data`	Executed Instructions $(\times 10^8)$: $5.8 \rightarrow 5.72$	27.6 (+11%)
Multiple iterations	Occupancy: $70.3\% \rightarrow 74.9\%$	27.4 (+0.5%)
Rows permutation	Occupancy: $74.9\% \rightarrow 81.8\%$	**27.1** (+1%)

8 Reference Software Implementation

For comparison purposes, we implemented SpMV on the three software instruction set architectures MMX, SSE and AVX, based on the RNS representation for the arithmetic and the CSR format for the storage of the matrix. We have not explored other formats that can be suitable for CPU. Probably blocked formats that better use the cache can further improve the performance on CPU. Unlike for GPU, processing separately the first dense columns accelerates the CPU SpMV of around 5%.

We can report the computational throughput in terms of GFLOP/s, which we determine by dividing the number of required operations (twice the number of non-zero elements in the matrix A multiplied by $2 \times n$) by the running time. We will use this unit of measure for the throughput even for integer instructions.

The experiment was run on an Intel CPU i5-4570 (3.2 GHz) using 4 threads on 4 cores. The AVX2 implementation using integers is the fastest implementation and reaches the highest throughput. However, the fact that the number of moduli is a multiple of four entail overheads. When comparing the software performance with the GPU one, the fastest software implementation is 4 to 5 times slower than on one graphics processor.

	Length of modulus	Number of moduli (n)	Timing in ms	Throughput in GFLOP/s
MMX (integer)	64	5	306	5.3
MMX (double-precision floats)	52	6	351	3.7
SSE2 (integer)	63	6	154	12.1
SSE2 (double-precision floats)	52	6	176	10.6
AVX2 (integer)	**63**	**8**	**117**	**21.2**
AVX2 (double-precision floats)	52	8	135	18.3
GPU (integer)	**64**	**5**	**27**	**57.6**

9   Conclusion

We have investigated different data structures to perform iterative SpMV for DLP matrices on GPUs. We have adapted the kernels for the context of large finite fields and added optimizations suitable to the sparsity and the specific computing model. The CSR-V kernel based on the *parallel* scheme appears to be the most efficient one. The SLCOO poses for the sizes that we use some hardware difficulties that nullify its contribution on increasing the cache hit rate. Future GPUs may enhance the performance. We have shown that using RNS for finite field arithmetic provides a considerable degree of independence, which can be exploited by massively parallel hardware. This implementation contributed to solving the discrete logarithm problem in $GF(2^{619})$ and $GF(2^{809})$ (See Appendix B and [3,10] for further details).

A   Formats and GPU Kernels of SpMV

$$\begin{pmatrix} 0 & a_{01} & 0 & a_{03} & 0 & 0 \\ 0 & a_{11} & 0 & 0 & a_{14} & a_{15} \\ a_{20} & 0 & a_{22} & a_{23} & 0 & 0 \\ 0 & a_{31} & 0 & 0 & a_{34} & 0 \\ 0 & a_{41} & a_{42} & 0 & 0 & a_{45} \\ 0 & 0 & a_{52} & 0 & 0 & a_{55} \end{pmatrix}$$

(a) Sparse matrix A

$$\text{data} = \begin{bmatrix} a_{01} & a_{03} & * \\ a_{11} & a_{14} & a_{15} \\ a_{20} & a_{22} & a_{23} \\ a_{31} & a_{34} & * \\ a_{41} & a_{42} & a_{45} \\ a_{52} & a_{55} & * \end{bmatrix} \quad \text{id} = \begin{bmatrix} 1 & 3 & * \\ 1 & 4 & 5 \\ 0 & 2 & 3 \\ 1 & 4 & * \\ 1 & 2 & 5 \\ 2 & 5 & * \end{bmatrix} \quad \text{len} = \begin{bmatrix} 2 & 3 & 3 & 2 & 3 & 2 \end{bmatrix}$$

(b) ELL-R representation

$$\text{data} = \begin{bmatrix} a_{01} & a_{11} & a_{03} & a_{14} & a_{15} & a_{20} & a_{31} & a_{22} & a_{23} & a_{34} & \cdots \end{bmatrix}$$

$$\text{row\_id} = \begin{bmatrix} 0 & 1 & 0 & 1 & 1 & 2 & 3 & 2 & 2 & 3 & \cdots \end{bmatrix}$$

$$\text{col\_id} = \begin{bmatrix} 1 & 1 & 3 & 4 & 5 & 0 & 1 & 2 & 3 & 4 & \cdots \end{bmatrix}$$

$$\text{ptrSlice} = \begin{bmatrix} 0 & 5 & 10 & \cdots \end{bmatrix}$$

(c) SLCOO-2 representation

Algorithm 5. ELL-R for row i executed by one thread

Inputs : **data**: array of $K \times N$ elements of K, **id**: array of $K \times N$ positive
integers,
len: array of N positive integers and u: vector of N elements of K.
Output: v: vector of N elements of K.

sum $\leftarrow 0$;
For $j \leftarrow 0$ to len$_i$ do
| $\quad$ sum $\leftarrow$ addmul(sum, data$_{N \times j + \text{row}}$, $u_{\text{id}_{N \times j + \text{row}}}$);
$v_i \leftarrow$ sum;

Algorithm 6. SLCOO-σ for slice i executed by thread of index tid in its warp

---

Inputs : **data**: array of n_{NZ} elements of K, **row_id**, **col_id**: arrays of n_{NZ}
positive integers, **ptrSlice**: array of N positive integers and u: vector
of N elements of K.
Output: v: vector of N elements of K.

Declare array sum $\leftarrow \{0\}$;   // array of σ elements of K
$i \leftarrow$ ptrSlice$_i$ + tid;   // position of beginning for each thread
While $j <$ ptrSlice$_{i+1}$ do
| $\quad$ sum$_{\text{row\_id}_j \bmod \sigma} \leftarrow$ addmul(sum$_{\text{row\_id}_j \bmod \sigma}$, data$_j$, $u_{\text{col\_id}_j \bmod \sigma}$);
| $\quad j \leftarrow j + 32$;
reduction_slcoo(sum, tid);   // reduction in *shared memory*
If tid= 0 then   // first thread of warp writes in *global memory*
| $\quad$ For $j \leftarrow 0$ to σ do
| $\quad\quad v_{i \times \sigma + j} \leftarrow$ sum$_j$;

B   Resolution of Linear Algebra of the Function Field Sieve

The linear algebra step consists of solving the system $Aw = 0$, where A is the matrix produced by the filtering step of the FFS algorithm. A is singular and square. Finding a vector of the kernel of the matrix is generally sufficient for the FFS algorithm.

The simple Wiedemann algorithm [27] which resolves such a system, is composed of three steps:

- *Scalar products:* It consists on the computation of a sequence of scalars $a_i = {}^t x A^i y$, where $0 \le i \le 2N$, and x and y are random vectors in $(\mathbb{Z}/\ell\mathbb{Z})^N$. We take x in the canonical basis, so that instead of performing a full dot product between ${}^t x$ and $A^i y$, we just store the element of $A^i y$ that corresponds to the non-zero coordinate of x.
- *Linear generator:* Using the Berlekamp-Massey algorithm, this step computes a linear generator of the a_i's. The output F is a polynomial whose coefficients lie in $\mathbb{Z}/\ell\mathbb{Z}$, and whose degree is very close to N.
- *Evaluation:* The last step computes $\sum_{i=0}^{deg(F)} A^i F_i y$, where F_i is the ith coefficient of F. The result is with high probability a non-zero vector of the kernel of A.

The Block Wiedemann algorithm [11] proposes to use m random vectors for x and n random vectors for y. The sequence of scalars is thus replaced by a sequence of $m \times n$ matrices and the numbers of iterations of the first and third steps become $(N/n + N/m)$ and N/n, respectively. The n subsequences can be computed independently and in parallel. So, the block Wiedemann method allows to distribute the computation without an additional overhead [25].

B.1   Linear Algebra of FFS for $GF(2^{619})$

The matrix has 650 k rows and columns. The prime ℓ is 217 bits. The computation was completed using the simple Wiedemann algorithm on a single NVIDIA GeForce GTX 580. The overall computation needed 16 GPU hours and 1 CPU hour.

B.2   Linear Algebra of FFS for $GF(2^{809})$

The matrix has 3.6M rows and columns. The prime ℓ is 202 bits. We run a Block Wiedemann on a cluster of 8 GPUs. We used 4 distinct nodes, each equipped with two NVIDIA Tesla M2050 graphics processors, and ran the Block Wiedemann algorithm with blocking parameters $m = 8$ and $n = 4$. The overall computation required 4.4 days in parallel on the 4 nodes.

These two computations were part of record-sized discrete logarithm computations in a prime-degree extension field [3,10].

References

1. Adleman, L.: A subexponential algorithm for the discrete logarithm problem with applications to cryptography. In: Proceedings of the 20th Annual Symposium on Foundations of Computer Science, Washington, DC, USA, pp. 55–60 (1979)
2. Bai, S., Bouvier, C., Filbois, A., Gaudry, P., Imbert, L., Kruppa, A., Morain, F., Thomé, E., Zimmermann, P.: Cado-nfs: Crible algébrique: Distribution, optimisation - number field sieve. http://cado-nfs.gforge.inria.fr/
3. Barbulescu, R., Bouvier, C., Detrey, J., Gaudry, P., Jeljeli, H., Thomé, E., Videau, M., Zimmermann, P.: Discrete logarithm in GF(2^{809}) with FFS. In: Krawczyk, H. (ed.) PKC 2014. LNCS, vol. 8383, pp. 221–238. Springer, Heidelberg (2014)
4. Bell, N., Garland, M.: Efficient sparse matrix-vector multiplication on CUDA. Technical report NVR-2008-004, NVIDIA Corporation, December 2008
5. Bell, N., Garland, M.: Cusp: Generic parallel algorithms for sparse matrix and graph computations (2012). http://code.google.com/p/cusp-library/
6. Bernstein, D.J.: Multidigit modular multiplication with the explicit chinese remainder theorem. Technical report (1995). http://cr.yp.to/papers/mmecrt.pdf
7. Blelloch, G.E., Heroux, M.A., Zagha, M.: Segmented operations for sparse matrix computation on vector multiprocessors. Technical report CMU-CS-93-173, School of Computer Science, Carnegie Mellon University, August 1993
8. Boyer, B., Dumas, J.G., Giorgi, P.: Exact sparse matrix-vector multiplication on GPU's and multicore architectures. CoRR abs/1004.3719 (2010)
9. Hayashi, T., Shimoyama, T., Shinohara, N., Takagi, T.: Breaking pairing-based cryptosystems using η_t pairing over GF(3^{97}). Cryptology ePrint Archive, Report 2012/345 (2012)
10. Jeljeli, H.: Resolution of linear algebra for the discrete logarithm problem using GPU and multi-core architectures. In: Silva, F., Dutra, I., Santos Costa, V. (eds.) Euro-Par 2014. LNCS, vol. 8632, pp. 764–775. Springer, Heidelberg (2014)
11. Kaltofen, E.: Analysis of coppersmith's block wiedemann algorithm for the parallel solution of sparse linear systems. Math. Comput. **64**(210), 777–806 (1995)
12. LaMacchia, B.A., Odlyzko, A.M.: Solving large sparse linear systems over finite fields. In: Menezes, A., Vanstone, S.A. (eds.) CRYPTO 1990. LNCS, vol. 537, pp. 109–133. Springer, Heidelberg (1991)
13. Lanczos, C.: Solution of systems of linear equations by minimized iterations. J. Res. Natl. Bur. Stand **49**, 33–53 (1952)
14. NVIDIA Corporation: CUDA Programming Guide Version 4.2 (2012). http://developer.nvidia.com/cuda-downloads
15. NVIDIA Corporation: PTX: Parallel Thread Execution ISA Version 3.0 (2012). http://developer.nvidia.com/cuda-downloads
16. Odlyzko, A.M.: Discrete logarithms in finite fields and their cryptographic significance. In: Beth, T., Cot, N., Ingemarsson, I. (eds.) EUROCRYPT 1984. LNCS, vol. 209, pp. 224–314. Springer, Heidelberg (1985)
17. Pollard, J.M.: A monte carlo method for factorization. BIT Numer. Math. **15**, 331–334 (1975)
18. Pomerance, C., Smith, J.W.: Reduction of huge, sparse matrices over finite fields via created catastrophes. Exp. Math. **1**, 89–94 (1992)
19. Schmidt, B., Aribowo, H., Dang, H.-V.: Iterative sparse matrix-vector multiplication for integer factorization on GPUs. In: Jeannot, E., Namyst, R., Roman, J. (eds.) Euro-Par 2011, Part II. LNCS, vol. 6853, pp. 413–424. Springer, Heidelberg (2011)

20. Sengupta, S., Harris, M., Zhang, Y., Owens, J.D.: Scan primitives for GPU computing, pp. 97–106, August 2007
21. Shanks, D.: Class number, a theory of factorization, and genera. In: 1969 Number Theory Institute (Proc. Sympos. Pure Math., Vol. XX, State Univ. New York, Stony Brook, N.Y., 1969), pp. 415–440. Providence, R.I. (1971)
22. Stach, P.: Optimizations to nfs linear algebra. In:CADO Workshop on Integer Factorization. http://cado.gforge.inria.fr/workshop/abstracts.html
23. Szabo, N.S., Tanaka, R.I.: Residue Arithmetic and Its Applications to Computer Technology. McGraw-Hill Book Company, New York (1967)
24. Taylor, F.J.: Residue arithmetic a tutorial with examples. Computer **17**, 50–62 (1984)
25. Thomé, E.: Subquadratic computation of vector generating polynomials and improvement of the block wiedemann algorithm. J. Symbolic Comput. **33**(5), 757–775 (2002)
26. Vázquez, F., Garzón, E.M., Martinez, J.A., Fernández, J.J.: The sparse matrix vector product on GPUs. Technical report, University of Almeria, June 2009
27. Wiedemann, D.H.: Solving sparse linear equations over finite fields. IEEE Trans. Inf. Theor. **32**(1), 54–62 (1986)

Finding Optimal Chudnovsky-Chudnovsky Multiplication Algorithms

Matthieu Rambaud[⊠]

Télécom ParisTech, 46 rue Barrault, 75013 Paris, France
matthieu.rambaud@telecom-paristech.fr

Abstract. The Chudnovsky-Chudnovsky method provides today's best known upper bounds on the bilinear complexity of multiplication in large extension of finite fields. It is grounded on interpolation on algebraic curves: we give a theoretical lower threshold for the smallest bounds that one can expect from this method (with exceptions). This threshold appears often reachable: we moreover provide an explicit method for this purpose.

We also provide new bounds for the multiplication in small-dimensional algebras over $\mathbf{F}_2$. Building on these ingredients, we:
- explain how far elliptic curves can provide upper bounds for the multiplication over $\mathbf{F}_2$;
- using these curves, improve the bounds for the multiplication in the NIST-size extensions of $\mathbf{F}_2$;
- thus, turning to curves of higher genus, further improve these bounds with the well known family of classical modular curves.

Although illustrated only over $\mathbf{F}_2$, the techniques introduced apply to all characteristics.

Keywords: Elliptic modular curves · Finite field arithmetic · Chudnovsky-Chudnovsky interpolation · Tensor rank · Optimal algorithms

1 Framework, Motivations and Goals

Let K be a field and $\mathcal{A}$ a finite-dimensional K-algebra[1]. The multiplication in $\mathcal{A}$, $m_{\mathcal{A}}$, is seen as a K-bilinear map:

$$m_{\mathcal{A}}: \qquad \mathcal{A} \times \mathcal{A} \longrightarrow \mathcal{A} \qquad\qquad (1)$$
$$(X, Y) \longrightarrow X.Y$$

Definition 1. The *symmetric bilinear complexity* of the multiplication $m_{\mathcal{A}}$ in $\mathcal{A}$, is the lowest integer n, such that there exists: n linear forms $(\phi_1, \ldots, \phi_n)$ on $\mathcal{A}$, along with n elements $(w_1, \ldots, w_n)$ of $\mathcal{A}$, such that $m_{\mathcal{A}}$ can be expressed as:

$$m_{\mathcal{A}}: (x, y) \longrightarrow \sum_{i=1}^{n} \phi_i(x).\phi_i(y).w_i \qquad\qquad (2)$$

---

[1] Which will here always be considered associative, commutative and unitary.

© Springer International Publishing Switzerland 2015
Ç. Koç et al. (Eds.): WAIFI 2014, LNCS 9061, pp. 45–60, 2015.
DOI: 10.1007/978-3-319-16277-5_3

This quantity[2] will be noted $\mu^{\mathrm{sym}}(\mathcal{A}/K)$.

Other complexity measures are possible, especially over the field $\mathbf{F}_2$. For example, one could count both the bitwise additions and multiplications. Or even take into account the possibility to perform computer-elementary operations on groups of 32 or 64-bits.

To compute the multiplication in large extensions of finite fields, the *interpolation method* of Chudnovsky and Chudnovsky [2], provides algorithms having the today's lowest known bilinear complexities. In our symmetric framework, the construction can be summarized as follows. Suppose one wants to compute the multiplication in $\mathbf{F}_{q^m}$ over $\mathbf{F}_q$. Start with an algebraic curve X over $\mathbf{F}_q$, equipped with a point Q of degree m, and convenient divisors D and G. For instance, let G be a collection of degree-1 points $P_1 \dots P_n$. Assuming some hypotheses are satisfied, the multiplication of any x and y in $\mathbf{F}_{q^m}$ can be performed with the following steps (numbered in the diagram below):

① *lift* x and y to some functions f_x and f_y, in the space of global sections $L(D)$, so that $f_x(Q) = x$ and $f_y(Q) = y$.

② *evaluate* f_x and f_y, separately, on each point P_i of the divisor G.

③ *compute*, for each P_i, the product of the two evaluations: $a_i = f_x(P_i).f_y(P_i)$. This is the critical step: here we perform $\deg G$ two-variables multiplications. We obtain the vector of values $(a_1, \cdots, a_n)$.

④ *interpolate* this vector to the *unique* function $g \in L(D+D)$ having values a_i at the P_i.

⑤ *evaluate* g at Q to find the product of x and y.

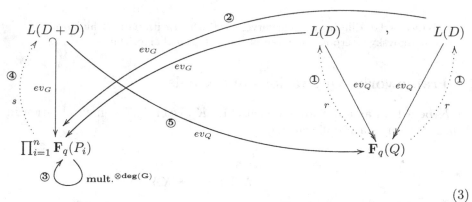

$$(3)$$

Such a triple (G, D, Q), in the precise sense of Theorem 2 below, is called an *interpolation system*. This theorem draws the consequences, of such an algorithm, on the symmetric bilinear complexity of multiplication in finite fields. But before stating it, some conventions must be done. A *curve* X will always be assumed projective, smooth and geometrically irreducible over $\mathbf{F}_q$, often $\mathbf{F}_2$. Given a

---

[2] Which is, in other words, the *partially-symmetric tensor rank* of $m_\mathcal{A}$, seen as an element of $(\mathcal{A}^* \otimes \mathcal{A}^*)^{\mathrm{Sym}} \otimes \mathcal{A}$ (this issue is addressed in general in [1]).

divisor D on X, the notation $\mathcal{O}(D)$ stands for the sheaf of sections of D. By $l(D)$ (resp. $i(D)$: the *index of specialty of* D) we will abridge the dimension of the $\mathbf{F}_q$-vector space of global sections $H^0(X, \mathcal{O}(D))$ (resp. of $H^1(X, \mathcal{O}(D))$). Recall that the Riemann-Roch theorem states: $l(D) - i(D) = \deg D + 1 - g$, where g is the genus of X.

More particularly, given P a closed point of X of degree n, D a divisor and l a positive integer, we will need the following map: the *evaluation of a global section of the line bundle* $\mathcal{O}(D)$ *at the thickened point* $P^{[l]}$, which takes values in $\frac{\mathbf{F}_{q^n}[y]}{y^l}$. Let t_P be a local parameter at P: multiplication by $t_P^{v_P(D)}$ provides a local trivialization of $\mathcal{O}(D)$ at P, and thus an evaluation map:

$$ev_Q : L(D) \longrightarrow \mathcal{O}_{X,P}/(t_P^l)$$
$$f \longrightarrow t_P^{v_P(D)} f_P \mod (t_P^l) \tag{4}$$

The target space maps isomorphically[3] to the $\mathbf{F}_q$-algebra $\frac{\mathbf{F}_q^n[y]}{y^l}$. This latter algebra has an internal multiplication map: note $\mu_q^{\mathrm{sym}}(m, l)$ for its symmetric bilinear complexity. $\mu_q^{\mathrm{sym}}(m, 1)$ is abbreviated as[4] $\mu_q^{\mathrm{sym}}(m)$.

Theorem 2 *([3] th 3.5).* Let X be a curve of genus g over $\mathbf{F}_q$, and let $m, l \geqslant 1$ be two integers. Suppose that X admits a closed point Q of degree $\deg Q = m$, and let t_Q be an uniformizing parameter. Let G be an effective divisor on X and write

$$G = u_1 P_1 + \cdots u_n P_n \tag{5}$$

where the P_i are pairwise distinct closed points, of degree $\deg P_i = d_i$ and resp. of uniformizing parameters t_{P_i}. Suppose there exists a divisor D on X such that:

(i) The product of evaluation maps at thickened closed points

$$L(D + D) \longrightarrow \prod_{i=1}^{n} \mathcal{O}_{X,P_i}/(t_{P_i}^{u_i})\mathcal{O}_{X,P_i} \tag{6}$$

is injective

(ii) The evaluation map

$$L(D) \longrightarrow \mathbf{F}_q(Q) \tag{7}$$

is surjective.

then

$$\mu_q^{\mathrm{sym}}(m) \leqslant \sum_{i=1}^{n} \mu_q^{\mathrm{sym}}(d_i, u_i) \tag{8}$$

---

[3] As shown in [3], Remark 3.4. This state-of-the-art version of the evaluation map on curves, is the key to the bounds recalled in Theorem 2.

[4] It is the symmetric complexity of the multiplication in the finite field extension $\mathbf{F}_{q^m}$.

Moreover,

(i') an equivalent condition for (i) is that $l(2D - G) = 0$,
(ii') while a sufficient condition for (ii) is that $i(D - Q) = 0$.

It is noticeable that, in the inequality 8, for a divisor G of fixed degree $\deg G$, written as $\sum_i u_i . \deg P_i$, the upper bound on $\mu_q^{\text{sym}}(m)$, given by the right-hand term of the last inequality, will be all the more large than the degrees of the points P_i and their multiplicities u_i are big. Thus, to minimise the symmetric bilinear complexity of the multiplication in $\mathbf{F}_{q^m}$, one is lead to follow this roadmap:

1 Collect (and improve) the best bounds for the $\mu_q^{\text{sym}}(m, l)$.
2 Find curves with many points P_i of low degree.
3 For such a curve X, fix a (small) degree $\deg G$ of G such that one hopes the existence of an admissible interpolation system (G, D, Q) on X (as precised in next section). For this candidate value of $\deg G$, find a combination $(u_i, P_i)_i$ of points and multiplicities that, numerically, minimises the upper bound of 8 under the constraint $\deg G \geqslant \sum_i u_i . \deg P_i$.
4 For this fixed candidate value $\deg G$, given such a numerically optimal $G = \sum_i u_i P_i$, check the existence of an interpolation system (G, D, Q).

1 is the motivation for the next paragraph. 2 is the motivation for the last section. 3 is an integer programme and will be illustrated in Sect. 3.2. 4 will be discussed in the next section.

2   New Bounds for Small-Dimensional Algebras $\frac{\mathbf{F}_{2^m}[y]}{y^l}$

Overview. In the Table 1 below are recapitulated the best known upper and lower bounds for the complexities $\mu_2^{\text{sym}}(m, l)$ of the multiplication in these algebras[5]. Each pair of lower-upper bound is given as "L-U". When no noticeable lower bound is known, the value of the lower bound L is left empty, so only "- U" is displayed. Finally, when the upper bound U is in fact optimal (so L=U), then one single value is displayed. In some cases, optimality results were unpublished to our knowledge. Therefore we detail our method of proof in the following paragraph. The three new upper bounds are displayed in bold[6] (new lower bounds beeing emphasized in Table 2).

In the following Table 2, we attempt to give references or explanations for some bounds. We do not claim to always giving credit to the first discoverer, nor to the most efficient method. In particular, the inequality $\mu_q^{\text{sym}}(m, l) \leqslant \mu_{q^d}^{\text{sym}}(e, l)\mu_q(d)$ (see *e.g.* [3] Lemma 4.6) is often used. For the upper and lower bounds that are new, up to our knowledge, we provide more details about how they were established. For place reasons, we refrain from displaying the formulas for the three new upper bounds here, but keep them available on demand.

---

[5] Note in particular that the first row shows the best known bounds for the multiplication in small finite field extensions of $\mathbf{F}_2$. On the opposite, the first column shows the best known bounds for the multiplication of polynomials modulo x^l over $\mathbf{F}_2$.

[6] The most noteworthy might be $\mu_2^{\text{sym}}(3, 2)$, because its (exact) value, 16, is strictly lower than the upper bound $18 = \mu_{2^3}^{\text{sym}}(1, 2)\mu_2^{\text{sym}}(3, 1)$.

Table 1. Lower−upper bounds on the complexities $\mu_2^{\text{sym}}(m, l)$

$l \backslash m$	1	2	3	4	5	6	7	8	9	10
1	1	3	6	9	13	15	16 − 22	−24	−30	−33
2	3	9	**16**	−24	.	.	.	.	.	.
3	5	15	−30	.	.	.	.	.	.	.
4	8	**−21**	.	.	.	.	.	.	.	.
5	11	.	.	.	.	.	.	.	.	.
6	14	.	.	.	.	.	.	.	.	.
7	16 − 18	.	.	.	.	.	.	.	.	.
8	−22	.	.	.	.	.	.	.	.	.
9	−27	.	.	.	.	.	.	.	.	.
10	**−30**	.	.	.	.	.	.	.	.	.

Table 2. Details for Table 1 entries

$\mu_2^{\text{sym}}(m, l)$	Upper bound	Lower bound
$(5, 1)$	[4]	[5]
$(6, 1)$	$\leqslant \mu_4^{\text{sym}}(3)\mu_2^{\text{sym}}(2)$ (first factor: by interpolation over $\mathbf{P}^1_{\mathbf{F}_4}$)	[5]
$(7, 1)$	[4]	[5]
$(8, 1)$	$\leqslant \mu_4^{\text{sym}}(4)\mu_2^{\text{sym}}(2)$ (unknown original contributor for the first)	.
$(9, 1)$	[6]	.
$(10, 1)$	$\leqslant \mu_4^{\text{sym}}(5)\mu_2^{\text{sym}}(2)$ (unknown original contributor for the first)	.
$(1, 5)$	[7]	[5]
$(1, 6)$	[7]	[5]
$(1, 7)$	[7] (proved valid over a general ring, in [8])	[5]
$(1, 8)$	[8]	[5]
$(1, 9)$	[8]	[5]
$(\mathbf{1}, \mathbf{10})$	$\leqslant \mu_4^{\text{sym}}(\mathbf{1}, \mathbf{5})\mu_2^{\text{sym}}(2)$, the first factor being *equal to 10*	.
$(2, 2)$	$\leqslant \mu_4^{\text{sym}}(1, 2)\mu_2^{\text{sym}}(2)$	**new**
$(2, 3)$	$\leqslant \mu_4^{\text{sym}}(1, 3)\mu_2^{\text{sym}}(2)$	**new**
$(\mathbf{3}, \mathbf{2})$	**new**	**new**
$(\mathbf{2}, \mathbf{4})$	$\leqslant \mu_4^{\text{sym}}(\mathbf{1}, \mathbf{4})\mu_2^{\text{sym}}(2)$, the first factor being *equal to 7*	.
$(4, 2)$	[3], inequality (94)	.
$(3, 3)$	$\mu_8^{\text{sym}}(1, 3)\mu_2(3)$	.

Update on Exhaustive Search Methods. To obtain the new upper and lower bounds, we built on the exhaustive search method introduced in [7], then in [5]. We would like first to share the techniques of implementation and search, that contributed to these results. And last, regarding the new lower bounds, we give the arguments that make our computational proofs valid and reproducible, especially when new shortcuts are involved.

Let K be a field, $\mathcal{A} = K^p$ a K-vector space of dimension p and B a (symmetric) K bilinear map, taking here values in $\mathcal{A}$, seen as a tensor in $\mathcal{A}^* \otimes \mathcal{A}^* \otimes \mathcal{A}$. Then, evaluation at the last component $\mathcal{A}$ defines a map[7]: $\mathcal{A}^* \to \mathcal{A}^* \otimes \mathcal{A}^*$, whose image is a K-vector subspace noted T. Let $\mathcal{G}$ be the set of (symmetric) bilinear forms of rank one in $\mathcal{A}^* \otimes \mathcal{A}^*$. Then the (partially symmetric) tensor-rank of B is equal to the least number k, of elements of $\mathcal{G}$, necessary to generate T. Going in the other direction, the incomplete basis theorem implies that: a subspace W of dimension k of $\mathcal{A}^* \otimes \mathcal{A}^*$, which both (i) is generated by elements of $\mathcal{G}$ and (ii) contains T, can be generated by a basis of T completed with elements of $\mathcal{G}$. These arguments validate the following algorithm ([5]), which both: given an integer k, determine if B is of rank strictly greater than k and, if not, find all the decompositions of length k of B.

Algorithm 3. Start with the subspace $W = T$ of $\mathcal{A}^* \otimes \mathcal{A}^*$, of dimension n. Then, for each element $\boldsymbol{g}$ of $\mathcal{G}$ independent from W, complete W by $\boldsymbol{g}$, to obtain $W' = W \oplus \boldsymbol{g}$ of dimension $n + 1$. Iterate until the dimension reaches k. Test if the subspace obtained is generated by elements of $\mathcal{G}$. If it is not the case for all the subspaces produced, it thus implies that the (partially symmetric) rank of B is strictly greater than k.

Here, we describe three implementation techniques that saved us significant computation time. (1) When looking for a symmetric decomposition of a symmetric bilinear form B, the entire research can be implemented in the subspace of symmetric bilinear forms. (2) As pointed by F. Courbier, the final step of the algorithm can be sped up. Instead of systematically computing the rank $\mathcal{G} \cap W$, one can check beforehand if its cardinality is lower than[8] $\dim W$. (3) To avoid testing several times the same subspace, one can fix once for all an ordering on $\mathcal{G} = (g_1, \ldots, g_M)$. Then, at each step of the recursion, complete $W = T \oplus K\boldsymbol{g}_{i_1} \oplus \ldots K\boldsymbol{g}_{i_s}$ by only the vectors $\boldsymbol{g}$ in $\mathcal{G}$ numbered *after* $\boldsymbol{g}_{i_s}$.

Finally, we put apart an observation that, either, helps finding quicklier a decomposition of given length k (and thus an upper bound), or, when none exists, gives a theoretical shortcut to establish this fact[9].

Observation 4. *Suppose that a group H of linear transformations of $E = \mathcal{A}^* \otimes \mathcal{A}^*$: (i) preserves the set $\mathcal{G}$ of symmetric rank-one bilinear forms, (ii) preserves the subspace T spanned by the components of the bilinear map B. Then, given an element g, there exists a subspace W of dimension k solution of the problem (i.e. (a) generated by rank one bilinear forms and (b) containing T), if and only if, for each element $h(g)$ in the orbit of g under H, there exists a subspace W' of dimension k which is a solution of the problem*

---

[7] This could be seen as a "tensor-flattening map", but we ignore how far this helps.

[8] This leads to noticing that, for algebras of dimension greater than 7, let k be the known upper bound for the tensor rank of multiplication, then a general subspace W of dimension k in $(\mathcal{A}^* \otimes \mathcal{A}^*)^{\mathrm{Sym}}$ will *a priori contain less than 0.01 rank-one tensor*. Thus, it would be interesting to know how to restrain the search to subspaces with a higher density of rank-one tensors.

[9] This method might be an elementary case of tensor decomposition methods. (It originated thanks to an apparently innocuous lecture of G. Cohen on cyclic codes.).

The *consequences* of this observation are easily illustrated in the case when one knows, in advance, a particular orbit $\mathcal{O}$ of rank-one symmetric bilinear forms, such that every possible solution subspace W, must at least contain one element of $\mathcal{O}$. One can then both reduce the time consumed: (1) by the search for an upper-bound. Indeed, one has the guarantee that, as soon as a solution of rank k exists, a recursive search of depth $k - n$ starting with any given element g_{i_0} in the orbit $\mathcal{O}$ will lead to a solution; (2) by the proof of a lower-bound: given any fixed element g_{i_0} in this orbit $\mathcal{O}$, if no subspace W containing g_0 is solution, then no solution subspace will contain any element in the orbit $\mathcal{O}$ of g_0, thus no solution exists. To have a proof of non-existence, one can thus restrain to a recursive search starting from g_{i_0}, hence divide the computation time by roughly half the cardinality of $\mathcal{A}$.

Here are two examples. In the case of a *finite field extension* $\mathbf{F}_{q^m}$ of $\mathbf{F}_q$, there is one single big orbit in the set $\mathcal{G}$ of symmetric bilinear forms, under the action of the group $H = \mathbf{F}_{q^m}^*$ defined by composition with two-side multiplication:

$$b \in \mathbf{F}_{q^m}^* : \lambda(\cdot\,,\cdot) \longrightarrow \lambda(b\cdot,b\cdot) \tag{9}$$

In the case of an *algebra* $\mathcal{A} = \frac{\mathbf{F}_{q^m}[y]}{y^l}$, there are l orbits $\{\mathcal{O}_1,\ldots,\mathcal{O}_l\}$ in $\mathcal{G}$ under the action of $\mathcal{A}^*$ (defined the same way as previously): $\mathcal{O}_j$ is the set of symmetric rank-one bilinear forms expressible as $\phi^{\otimes 2}$, where ϕ sends to zero every polynomial of $\frac{\mathbf{F}_{q^m}[y]}{y^l}$ of valuation in y greater or equal to j. In particular the last orbit $\mathcal{O}_l$, which is the largest, consists of elements $\phi^{\otimes 2}$, for which ϕ is not zero on at least one polynomial of valuation $l - 1$.

Regarding the last example, one remarks, in addition, that any minimal (symmetric) multiplication algorithm will involve at least one element of the greatest orbit $\mathcal{O}_l$. Hence, the previously discussed consequences apply. In particular, (2) underlies our computational proofs for lower bounds. This computations were performed with the C library dedicated to fast linear algebra in characteristic 2, [9], But far less computer resources were used than in [5], so we hope that these refinements of the method will help further improvements.

3   Best Expectable Complexity Using a Given Curve

3.1   Best Expectable Interpolation Systems

Let X be a curve of genus g over $\mathbf{F}_q$. An *optimal symmetric interpolation system on X with respect to multiplication in a finite field extension* $\mathbf{F}_{q^m}$, is a triple (G, D, Q) that provides a symmetric multiplication algorithm in $\mathbf{F}_{q^m}$, reaching a *lower bound* for the complexity of the Chudnovsky-Chudnovsky interpolation method on X. This will be precised in Definition 9.

The first two key-observations were brought to us[10] by Randriambololona.

---

[10] We do not claim either to having first drawn the consequences which follow.

Observation 5. When $H^1(X, \mathcal{O}(D)) = 0$, the sufficient condition (ii') in Theorem 2 is in fact equivalent to (ii). Moreover, it is remarkable that this situation happens, for instance, when $\deg D \geq 2g - 2$, by the Riemann-Roch theorem.

Indeed, one then has the shortened exact sequence:

$$0 \to H^0(X, \mathcal{O}(D - Q)) \to H^0(X, \mathcal{O}(D)) \xrightarrow{ev_Q} \mathcal{O}_{X,Q}/(t_Q) \to \dots$$
$$\dots \to H^1(X, \mathcal{O}(D - Q)) \to 0 \tag{10}$$

Observation 6. Let X be a curve of genus g over $\mathbf{F}_q$, m an integer and Q a closed point of degree m. Suppose that there exists an interpolation system (G, D, Q) on X, with Q of degree m, which furthermore satisfies the sufficient condition (ii') of Theorem 2. Then[11], $\deg G \geqslant 2m + g - 1$.

Proof. Let (G, D, Q) be the interpolation system of the hypothesis, and n (reps. d) the degree of G (resp. D). By condition (i) of Theorem 2, $l(2D - G) = 0$. Thus, the Riemann-Roch theorem applied to $2D - G$ implies:

$$2d - n \leqslant g - 1 \tag{11}$$

In addition, by condition (ii') of Theorem 2, which is satisfied here by assumption, $i(D - Q) = 0$. Thus, the Riemann-Roch theorem applied to $D - Q$ implies $d - m \geqslant g - 1$. Multiplication by -2 of this inequality leads:

$$-2(d - m) \leqslant -2(g - 1) \tag{12}$$

Summing (11) with (12), leads to $2m - n \leqslant -g + 1$. $\qquad\square$

Lemma 7. Let X be a curve of genus g, m an integer and Q a closed point of degree m. Suppose furthermore that $\boldsymbol{m > g}$. Then, any divisor D belonging to an interpolation system (G, D, Q) on X satisfies $\deg D \geqslant m + g - 1$.

Proof. Note d the degree of D.

First case: suppose $2g - 2 < d < m + g - 1$. Then, for degree reasons, $i(D) = 0$. Thus the Riemann-Roch theorem implies $l(D) = d + 1 - g < m = \dim \mathbf{F}_q(Q)$. Therefore, the evaluation map (ii) of Theorem 2 cannot be surjective, for dimension reasons.

Last case: suppose $d \leqslant 2g - 2 < m + g - 1$. Then, K being the canonical divisor of X, the Riemann-Roch theorem and Serre duality imply that $l(D) = l(K - D) + d + 1 - g$. But $K - D$ being of non-negative degree, we also have the bounding $l(K - D) \leqslant 2g - 2 - d + 1$. Thus, $l(D) \leqslant g < m = \dim \mathbf{F}_q(Q)$, thus the same contradiction as previously. $\qquad\square$

---

[11] The following inequality bounds below the complexity of a multiplication algorithm by interpolation on a given curve. Indeed, recall that the degree of G, bounds below this complexity. This bound is more constraining than the general lower bound, $2m - 1$, for the bilinear rank of multiplication in integral algebras (*cf.* [3] Lemma 1.9). Indeed, one considers here only a particular category of multiplication algorithms: the interpolation on curves method (here, the sub-category satisfying (ii'), but this restriction will be lifted, as soon as $g < m$).

Proposition 8 *(Properties of optimal interpolation systems).* Let X be a curve of genus g, m an integer and Q a closed point of degree m. Suppose furthermore, as previously, that $m > g$. Then:

1. The degree of an interpolation divisor G, belonging to an interpolation system (G, D, Q), cannot be lower than[12] $2m + g - 1$;
2. For such an interpolation system, i.e. with $\deg G$ attaining the lower bound $2m + g - 1$, then the degree of D is necessarily equal to $m + g - 1$.

Proof. Let (G, D, Q) be an interpolation system with Q of degree m. Note d the degree of D. Firstly, d being strictly greater than $2g - 2$, Observation 5 applies. Hence, the interpolation system satisfies (ii'). Thus, Observation 6 applies: G cannot be of degree lower than $2m + g - 1$.

For the second part, let (G, D, Q) be an interpolation system as in the assumption, that is with Q of degree m and $\deg G$ attaining the lower bound $2m + g - 1$. Then, recall that by inequality (11), $2d - \deg G \leqslant g - 1$. Thus here, $d \leqslant m + g - 1$. But by the previous lemma, we also have the opposite inequality: $d \geqslant m + g - 1$. □

Definition 9. Let X be a curve of genus g over $\mathbf{F}_q$, and m an integer. Suppose furthermore, as previously, that $m > g$. An *optimal interpolation system on X in degree m* is a triple (G, Q, D), with Q of degree m, that satisfies the three following conditions:

1. Satisfies the conditions (i') and (ii') of Theorem 2;
2. $\deg G$ reaches the lower bound $2m + g - 1$ of Proposition 8;
3. G is *numerically optimal*, that is: write $G = u_1 P_1 + \cdots u_n P_n$, then, this combination of points P_i and multiplicities u_i minimizes the upper bound on $\mu_q^{\mathrm{sym}}(m)$ given by Theorem 8.

Proposition 10 *(Effective construction).* Let X be a curve of genus g, such that there exists an optimal interpolation system (G, D, Q), with Q of degree m. Note $\mathrm{Cl}^0(X)$ the zero-class group of X. Then, it is possible to build (G, D, Q) with at most twice $\#\mathrm{Cl}^0(X)$ emptiness-checkings of Riemann-Roch spaces[13].

Indeed, first notice that, (G, D, Q) being optimal by assumption, the degree of D is $m+g-1$ by Proposition 8. In particular by 5, the sufficient condition (ii') of Theorem 2 is actually *equivalent* to (ii). Thus, one does not miss any optimal interpolation system (G, D, Q) by checking conditions (i') and (ii') instead of (i) and (ii). Secondly, notice that conditions (i') and (ii') depend only on the class of $D - Q$ (resp. $2D - G$) in $\mathrm{Cl}^0(X)$.

---

[12] It is to be noted that today's best uniform and asymptotic bounds, over $\mathbf{F}_2$, are obtained: (1) with curves of genus $g \geqslant m$ (one gets $g \approx 2.5m$ from the proof of [10] Theorem 4.1) (2) but with degrees of G still greater than $2m + g - 1$. It would thus be interesting to know if one can improve this threshold.

[13] Actually, both computations for (i') ("Step 1") and (ii') ("Step 2") will occur, here, in Cl^{g-1}, so one can remove the factor 2, as soon as one keeps in memory all the classes of divisors already tested.

Step 1: Look for a numerically optimal G, of degree $2m + g - 1$, whose class has not been already produced in the previous runs[14] of Step 1, then proceed to Step 2.

– Step 2: look for a divisor D, of degree $m + g - 1$, such that $l(2D - G) = 0$, and such that the class of D has not been considered yet in the previous runs[15] of Step 2. If such a D exists, proceed to Step 3.
  • Step 3: find every possible closed point Q of degree m, such that the class of $D - Q$ in $\mathrm{Cl}^{g-1}(X)$ has not been tested yet in the previous runs of Step 3, and then test if $i(D - Q) = 0$ [16]. If so, return[17] (G, D, Q).
  • If we are here, this means that the last run of Step 3 did not return any solution. Assuming that an optimal interpolation system does exist, this implies that there remain classes $(C_1 \ldots C_s)$ in $\mathrm{Cl}^m(X)$, which have not been tested yet in the previous runs of Step 3. Thus, return to Step 2.
– If we are here, it means that no divisors D were found in Step 2. Then, the assumption for the existence of an interpolation system implies that: there exists another numerically optimal divisor G'', and another D'', such that there exist classes $(C_1, \cdots, C_s)$ in $\mathrm{Cl}^{g-1}(X)$, that have not been tested in Step 3 and are of the form $(D'' - Q_i)_{i \in I}$. Thus, return to Step 1.

---

[14] Enumerating the (classes of) *numerically optimal* divisors on X is performed in two steps: (1) enumerate each collections of integers $(n_{d,u})_{d,u}$ (where $n_{d,u}$ stands for the number of points of degree d involved with multiplicity u in G), that (a) minimise the upper bound of Theorem 8: $\sum_{d,u} n_{d,u} \mu_q^{\mathrm{sym}}(d, u)$, under the constraints that (b) the total degree $\sum_{d,u} n_{d,u} du$ (is greater or) equal to the above lower bound $2m + g - 1$, and (c) for each d, $\sum_u n_{d,u}$ is lower or equal to the number of points of degree d in X. (2) for each collection $(n_{d,u})_{d,u}$, enumerate the divisors involving exactly $n_{d,u}$ points of degree d with multiplicity u.

[15] This involves at most $\#\mathrm{Cl}^0(X)$ Riemann-Roch spaces emptiness checkings in $\mathrm{Cl}^{g-1}(X)$ (minus those already performed in the previous runs).

[16] This involves at most $\#\mathrm{Cl}^0(X)$ Riemann-Roch spaces checkings in Cl^{g-1}. (Notice here that, $D - Q$ being of degree $g - 1$, the Riemann-Roch theorem implies that this condition is equivalent to $l(D - Q) = 0$).

[17] Under the additional assumption where points Q of degree m would exist in every single class $\mathrm{Cl}^m(X)$, then the first run of Step 3 always returns a solution as soon as an optimal interpolation system exists. Thus, if no solution is returned, this is a proof that no optimal interpolation system of degree m does exist on X. It is to be noted that, even if the case of genus one curves can be treated directly, a proof of the assumption in this case does exist. More precisely, [11] Theorem 27 states that, for $q \geqslant 7$ (and presumably $\geqslant 4$ for m sufficiently large), for $m \leqslant 2^{4096}$, there exists a prime divisor of degree m in every class. Any analogous proof in higher genus would be of interest. [There is actually a mistake in Lem. 19: in the first line of (2), μ is actually meant to be n/v_r. Thus, in the last but one line, μ can actually be equal to 1 when n has no square factors. This is compensated when, *e.g.*, m is greater than $6! = 720$].

3.2 The Example of Elliptic Curves, over $\mathbf{F_2}$

Lowest Expectable Value for the Degree of G. We first recall the known sufficient conditions to build interpolation algorithms on elliptic curves over a general $\mathbf{F}_q$ ([3], Proposition 4.3):

Proposition 11. *Let X be an elliptic curve over $\mathbf{F}_q$, with all notations as in Theorem 2, and P_∞ the zero element of the group of points of $X(\mathbf{F}_q)$. Let m be an integer. Suppose that X admits a closed point Q of degree[18] m. Let G be an effective divisor on X, and write:*

$$G = u_1 P_1 + \ldots + u_n P_n \tag{13}$$

where the P_i are pairwise distinct closed points, of degree $\deg P_i = d_i$, so $\deg G = \sum_{i=1}^n d_i u_i$. Then,

$$\mu_q^{\mathrm{sym}}(m) \leqslant \sum_{i=1}^n \mu_q^{\mathrm{sym}}(d_i, u_i)$$

provided one of the following conditions is satisfied:

(i) $\deg G = 2ml$ *and* $|X(\mathbf{F}_q)| \geqslant 2$ *and* $\mathrm{Cl}^0(X)$ *is not entirely of 2-torsion.*
(ii) $|X(\mathbf{F}_q)| \geqslant 2$ *and* $\deg G \geqslant 2ml + 1$. *Furthermore, if the additional criterion on G is satisfied: the divisor*

$$G - \deg G.P_\infty \text{ is not equivalent to } 0$$

then $\deg G$ can even be taken equal to $2ml$.
(iii) $\deg G \geqslant 2ml + 3$.

Taking the example of $\mathbf{F}_2$, the five equivalence classes of elliptic curves over this base field are given by the following equations.

$$y^2 + y + x^3 + x + 1 = 0 \tag{14}$$
$$y^2 + xy + x^3 + x^2 + 1 = 0 \tag{15}$$
$$y^2 + y + x^3 = 0 \tag{16a}$$
$$y^2 + y + x^3 + x = 0 \tag{16b}$$
$$y^2 + xy + x^3 + 1 = 0 \tag{16c}$$

In Table 3 below, we classify these curves along the previous conditions. For each curve X we give: the number B_1 of integral points, the structure of the group of points $\mathrm{Cl}^0(X)$ and, thus, the smallest degree of G satisfying the previous sufficient conditions: we call this value "upper-bound". In addition, we also bound below the degree $\deg G$ of an interpolation divisor on X (distinguishing whether or not the sub-condition on G is satisfied). Regarding the 2-torsion case, we finally explain why it is in fact nearly always possible to find a divisor

---

[18] A sufficient condition for that is $m \geqslant 7$ *cf.* for example [12] V.2.10 c).

Table 3. Lower-upper bounds for the degree of the best interpolation divisor G

Curve	B_1	Cl^0	Additional criterion on G	Lower bound on $\deg G$	Upper bound on $\deg G$	Is the additional criterion on G satisfiable?
(14)	1	0	.	$2ml + 3$	$2ml + 3$	.
(15)	2	$\mathbf{Z}/2$	when false:	$2ml + 1$	$2ml + 1$	**yes, for nearly all $m \leqslant 2^{4096}$**
			when true:	$2ml$	$2ml$	
(16a)	3	$\mathbf{Z}/3$	.	$2ml$	$2ml$	.
(16b)	5	$\mathbf{Z}/5$				
(16c)	4	$\mathbf{Z}/4$				

G satisfying the subcondition. Before giving proofs, we can notice that all the lower bounds were actually reached by the known upper-bounds[19].

Demonstrations **Curves** (16a), (16b), (16c) **and [(15) when condition on G true]:** the *lower-bounds* settled at $2ml$ are a direct consequence of Proposition 8,1[20].

Curve (15) **- when condition on G false:** if $\deg G$ were $2m$, then by Proposition 8,2 the degree of D would by m, thus $2D - G$ would be in the zero-class, thus $l(2D - G)$ would be one, which contradicts (i) by Riemann-Roch.

Curve(14): *Firstly*, it is not possible to build a degree $2m$ interpolation divisor G. We reuse the arguments of [3] 4.7. The evaluation map $ev_Q \colon L(D) \longrightarrow \mathcal{O}_{X,Q}/(t_Q)$ fits in the long exact sequence

$$0 \to H^0(X, \mathcal{O}(D - Q)) \to H^0(X, \mathcal{O}(D)) \xrightarrow{ev_Q} \mathcal{O}_{X,Q}/(t_Q) \to \dots$$
$$\dots \to H^1(X, \mathcal{O}(D - Q)) \to 0 \quad (17)$$

But, D being of degree m by Proposition 8, the Riemann-Roch theorem implies that $l(D) = m$. Also, the divisor $D - Q$ having degree 0 and (iii) having trivial class group, $D - Q$ is then equivalent to zero, thus $l(D - Q)$ is equal to 1. As a result, dimension-counting implies that the evaluation map ev_Q has image of dimension lower or equal to $m - 1$, thus cannot be surjective.

Secondly, $\deg G$ cannot be equal to $2m + 1$. Indeed, the previous arguments shows that, in order to have the surjectivity of the evaluation at Q map, one must have $d = \deg D > m$. Write $d = m + i$. Then, $\deg(2D - G) = 2i - 1 > 0$. Thus, $i(2D - G) = 0$ for degree reasons. So, by the Riemann-Roch theorem, $l(2D - G) = 2i - g = 2i - 1 > 0$. So the condition (ii') is false. But recall that,

---

[19] The proofs and results for this column are the same on a general base field $\mathbf{F}_q$. And regarding the discussion on the divisor G for the full 2-torsion curve (15), such cases of curves arise in finite number (indeed, it is a basic fact that the 2-torsion group of an elliptic curve is included in $\mathbf{Z}/2\mathbf{Z} \times \mathbf{Z}/2\mathbf{Z}$, and on the other hand, curves have enough points for q sufficiently large). Furthermore, the classification provided by [13] shows that this number is small.

[20] And were probably known since Shokrollahi 1992.

by Observation 5, the degree of D being greater than $2g - 2$, condition (ii) of Theorem 2 is also not satisfied.

Finally, deg G cannot be equal to $2m+2$. For that, writing again $d = m+i > m$, two cases are possible:

- Either $i > 1$, thus $\deg(2D - G) = 2i - 2 > 0$ so by the same argument as above, condition (ii) of Theorem 2 is not satisfied;
- Or, $i = 1$. But then $\deg(2D - G) = 0$, thus linearly equivalent to zero, by triviality of the zero-class group of curve (iii). Thus $l(2D - G) = 1$, contradicting condition (i) of 2.

Curve (15) - satisfiability of the condition on G: *Claim*[21]: for each degree $m \leqslant 2^{4096}$, there exists a point P of degree m such that $P - mP_\infty$ does not lie in the zero class. As a consequence, for nearly all m, given a numerically optimal divisor G of degree $2m$ on Curve (15) (furthermore assumed not built using points of degree greater than 2^{4096}), it is possible to deduce a numerically optimal G' that does not lie in the zero-class (by swapping a couple of points and/or multiplicities)[22].

New Bounds over NIST-Size Extensions of $\mathbf{F_2}$. Five finite field extensions of $\mathbf{F_2}$ are recommended by the NIST in [14] to perform elliptic-curve based cryptography, of degrees from $m = 163$ to 571. The best known bounds for the symmetric complexities of the multiplication in these extensions have been set in [13]. To achieve this, the authors used interpolation divisors G of the smallest degree given by the previous sufficient conditions 11. But we have just shown, in the previous paragraph, that these conditions on deg G could not be sharpened. Nevertheless, it is still possible to improve these five bounds.

Firstly, the authors seem to have used the value 15 as upper-bound on $\mu_2^{\mathrm{sym}}(1, 6)$, although a better value, 14, is known (*cf.* Table 1). Using this value, and interpolating on the curve of Eq. (16b) instead of (16a), already provides better bounds for all the five extensions considered. This is shown below in the third column of Table 4. Secondly, plugging in the three new bounds given in Table 1, leads to further improved bounds for the two last extension degrees, as shown below in the last column of Table 4.

4   Further Improvements, with Classical Modular Curves

4.1   Method

The well-known family of classical modular curves $X_0(N)_{\mathbf{F}_p}$ is a natural candidate to build interpolation systems. Indeed, they have been noticed for their

---

[21] For the proof: one adapts the estimations in [11] that lead to Theorem 16 (1), paying attention to a small mistake in the proof (see footnote 17), replacing q and the p_i-torsion by their values, taking m great enough to compensate the new positive terms, and computationally check the values of m below this threshold.

[22] Indeed, the possible degrees deg $G = m_i$ for which this swapping is not possible, lie among those for which all the points P_i of X -up to a certain degree n_i- occur in G with equal multiplicities. Therefore, the gaps in the sequence of the excluded degrees $(m_i)_i$ take values in the (growing) set $\{B_j, j \in \{1, \ldots, n_i\}\}$.

Table 4. Today's best possible upper bounds for $\mu_2^{\mathrm{sym}}(m)$ using elliptic curves

m	Former upper bound ([13])	Better bounds - with existing ingredients	Best bounds - using improved ingredients
163	906	**905**	905
233	1340	**1339**	1339
283	1668	**1661**	1661
409	2495	**2494**	**2492**
571	3566	**3563**	**3562**

high asymptotical rate of integer points of degree 2 (which led to improve a famous bound in coding theory). But also, computational tools exist to study them from different angles, as emphasized by what follows:

To begin, the results of Igusa show that (1) for all positive integer N, there exists an integral model $X_0(N)_{\mathbf{Z}}$ of the modular curve $X_0(N)_{\mathbf{Q}}$ having good reduction over all the primes $p \nmid N$. (2) Moreover, the Hecke correspondences T_{p^i}, acting on $\mathrm{Pic}^0(X_0(N)_{\mathbf{Q}})$, can be reduced over these primes p to correspondences $\widetilde{T}_{p^i}$, in a way compatible with the moduli interpretation (both results stated *e.g.* in [15], Theorem 8.6.1, resp. Diag. 8.30). Then, the Eichler-Shimura congruence relation implies that the number of $\mathbf{F}_{p^m}$-points in $X_0(N)_{\mathbf{F}_p}$ can be computed from the value of the trace of the Hecke correspondances $\widetilde{T}_{p^i}$, acting on $\mathrm{Pic}^0(X_0(N)_{\mathbf{F}_p})$. An explicit formula is provided in [16], Cor. 5.10.1. As a result, this enables to count closed points in the modular curves $X_0(N)_{\mathbf{F}_p}$ with analytic tools. Indeed, $\mathrm{Pic}^0(X_0(N)_{\mathbf{C}})$ is dual to the space of holomorphic forms $H^0(\Omega_{X_0(N)}, \mathbf{C})$.

But computing with cusp-forms expansions cannot be performed in large level N. Instead, the previous space is preferably seen in the (twice) larger space of complex differential forms on $X_0(N)_{\mathbf{C}}$. Indeed one can describe, in a purely algebraic fashion, the action of the Hecke operators on the Poincaré dual, $H_1(X_0(N), \mathbf{C})$, using a preferred basis called "Manin symbols". This action is implemented in Sage [17]. Then, to retrieve the trace of the Hecke operators on the subspace, $H^0(\Omega_{X_0(N)}, \mathbf{C})^*$, one can show that it suffices to divide by two the total trace[23].

Having computed, with the previous method, the number of closed points of degrees up to 10 on the $X_0(N)_{\mathbf{F}_2}$ for N up to 1300, one can select those which provide the best numerically optimal divisors G, for the five extension of

---

[23] Indeed, there exists a basis of holomorphic forms $H^0(\Omega_{X_0(N)}, \mathbf{C})$ such that the Hecke operators T_n act by matrices with coefficients in $\mathbf{Q}$. An elementary way to see this is to consider the $\mathbf{Q}$-algebra $\mathbf{T}' \subset \mathrm{End}(H^0(\Omega_{X_0(N)}, \mathbf{C}))$ generated by the Hecke operators acting on holomorphic forms. Then, noting $\mathbf{T}'^* = \mathrm{Hom}_{\mathbf{Q}}(\mathbf{T}', \mathbf{Q})$ the dual algebra, and extending the scalars by $\mathbf{C}$, Proposition 3.24 of [18] establishes a natural isomorphism between $\mathbf{T}'^*_{\mathbf{C}}$ and $H^0(\Omega_{X_0(N)}, \mathbf{C})$. To conclude, one can check that this isomorphism transports the natural action of $\mathbf{T}'$ on $\mathbf{T}'^*$, when extended over $\mathbf{C}$, to the action of $\mathbf{T}'$ on $H^0(\Omega_{X_0(N)}, \mathbf{C})$.

$\mathbf{F}_2$ considered. One can then compute rational projective models of some low genus curves, using the canonical embedding method developed in [19][24]. Notice, at this stage, that one has no guarantee that the models have good reduction modulo 2. This is often not the case, which explains the empty fifth column of the following Table 5. It finally remains to check that these divisors G do belong to an optimal interpolation system, using the construction described in 10. This can be done in a timely manner, using a well-known proprietary software ([21]), which implements an algorithm of Hess for Riemann-Roch spaces computations.

4.2 Results

The following Table 5 gives the best bounds obtained, using curves up to genus 6, for the $X_0(N)_{\mathbf{F}_2}$ which could actually be computed.

Let us describe a reproducible run of Algorithm 10, leading to the best entry (in bold) 900, for the extension degree $m = 163$. It is performed on the genus 6 curve $X = X_0(71)_{\mathbf{F}_2}$. The lowest expectable degree for G is $2 * 163 + 6 - 1 = 331$ and, in this case, D should be of degree $163 + 6 - 1 = 168$. Setting the random seed to 0 in Magma, we fix once for all an enumeration of the points of X (up to degree 8)[25], and an isomorphism of the class group of X with $\mathbf{Z}/315\mathbf{Z} \oplus \mathbf{Z}$: the first generator (of degree 1) being called D_1 and the second (of degree 0), D_2.

Step 1: (using the notations of Footnote 14) (1) (a) fix an optimal collection of integers: $n_{1,5} = 1$, $n_{1,6} = 3$, $n_{2,4} = 3$, $n_{3,1} = 4$, $n_{4,1} = 6$, $n_{5,1} = 4$, $n_{6,1} = 10$, $n_{8,1} = 21$, (b) which is of total degree 331 (c) and is compatible with the number of points on X of respective degrees up to 8. (2) Build a divisor G from this collection: a first attempt is to use the points in the order they were enumerated (so that, for the four points of degree one on X, the first is given multiplicity 5 and the three remaining multiplicity 6).

– Step 2 Building the class of D as $i.D_1 + 168.D_2$, with a varying coefficient i for D_1, it happens that for $i = 2$, the condition $l(2D - G) = 0$ is satisfied.

Table 5. Upper bounds on $\mu_2^{\mathrm{sym}}(m)$, sorted by the genus of the curves used

$m\backslash g$	1 (Table 4)	2	3	4	5	6
163	905		903	901		.	**900**
233	1339		1336	.	1335	.	.
283	1661		1660	.	1654	.	.
409	2492		2491	.	2486	.	.
571	3562		3561	3560	3555	.	.

---

[24] For the genus 4 hyperelliptic curve $X_0(47)_{\mathbf{Q}}$, the plane integral model provided in [20], appeared to have good reduction over 2.

[25] It then results from (8) that a numerically optimal G would lead to 900.

- Step 3: With various random seeds, we generate random points Q of degree m in several classes[26]. It happens that for seed 1, $i(D - Q) = 0$, thus giving an optimal interpolation system (G, D, Q).

References

1. Bernardi, A., Brachat, J., Comon, P., Mourrain, B.: General tensor decomposition, moment matrices and applications. J. Symbolic Comput. **52**, 51–71 (2013)
2. Chudnovsky, D., Chudnovsky, G.V.: Algebraic complexities and algebraic curves over finite fields. J. Complex. **4**, 285–316 (1988)
3. Randriambololona, H.: Bilinear complexity of algebras and the Chudnovsky-Chudnovsky interpolation method. J. Complex. **28**, 489–517 (2012)
4. Montgomery, P.L.: Five, six and seven-term Karatsuba-like formulae. IEEE Trans. Comput. **54**, 362–370 (2005)
5. Barbulescu, R., Detrey, J., Estibals, N., Zimmermann, P.: Finding optimal formulae for bilinear maps. In: Özbudak, F., Rodríguez-Henríquez, F. (eds.) WAIFI 2012. LNCS, vol. 7369, pp. 168–186. Springer, Heidelberg (2012)
6. Cenk, M., Özbudak, F.: Improved polynomial multiplication formulas over $\mathbf{F}_2$ using CRT. IEEE Trans. Comput.- Brief Contributions **58**, 572–577 (2009)
7. Oceledets, I.: Optimal Karatsuba-like formulae for certain bilinear forms in GF(2). Linear Algebra Appl. **429**, 2052–2066 (2008)
8. Cenk, M., Özbudak, F.: Multiplication of polynomials modulo x^n. Theoret. Comput. Sci. **412**, 3451–3462 (2011)
9. Albrecht, M.: The M4rie library for dense linear algebra over small fields with even characteristic. Arxiv 1111.6900 (2011)
10. Pieltant, J., Randriambololona, H.: New uniform and asymptotic upper bounds on the tensor rank of multiplication in extensions of finite fields (2013)
11. Shokrollahi, M.A.: Counting prime divisors on elliptic curves and multiplication in finite fields. In: Joyner, D. (ed.) Coding theory and Cryptography, pp. 180–201. Springer, Heidelberg (2000)
12. Stichtenoth, H.: Algebraic Function Fields and Codes. Springer, Heidelberg (1993)
13. Ballet, S., Bonnecaze, A., Tukumuli, M.: On the construction of Chudnovsky-type algorithms for multiplication in large extensions of finite fields (2013)
14. NIST: FIPS 186-4 (2013)
15. Diamond, F., Shurman, J.: A First Course in Modular Forms. Springer, New York (2004)
16. Moreno, C.J.: Algebraic Curves on Finite Fields. Cambridge University Press, Cambridge (1993)
17. Stein, W., et al.: Sage mathematics software (Version 6.3). The Sage development team (2014). http://www.sagemath.org
18. Stein, W.: Modular Forms, a Computational Approach. AMS, Providence (2006)
19. Galbraith, S.D.: Equations For Modular Curves. Ph.D. Thesis, Oxford (1996)
20. Yang, Y.: Defining equations of modular curves. Adv. Math. **204**, 481–508 (2006)
21. Bosma, W., Cannon, J., Playoust, C.: The magma algebra system. i. the user language. J. Symbolic Comput. **24**, 235–265 (1997)

---

[26] This is currently achieved by splitting only the place at infinity, so we do not know if this leads to every possible class for points of degree m.

Reducing the Complexity of Normal Basis Multiplication

Ömer Eğecioğlu and Çetin Kaya Koç[✉]

Department of Computer Science, University of California Santa Barbara,
Santa Barbara, USA
{omer,koc}@cs.ucsb.edu

Abstract. In this paper we introduce a new transformation method and a multiplication algorithm for multiplying the elements of the field $GF(2^k)$ expressed in a normal basis. The number of XOR gates for the proposed multiplication algorithm is fewer than that of the optimal normal basis multiplication, not taking into account the cost of forward and backward transformations. The algorithm is more suitable for applications in which tens or hundreds of field multiplications are performed before needing to transform the results back.

1 Introduction

Arithmetic operations in finite fields $GF(q)$ have several applications in cryptography, coding, and computer algebra. Particularly of interest are fields of characteristic 2, where $q = 2^k$, which have various uses in elliptic curve cryptography for large values of k, usually in the range from 160 to 521. Furthermore, smaller fields, for example, $k = 8$ (AES/Rijndael) and $k = 2, \ldots, 32$ (Reed-Solomon and BCH codes) are also commonly used. Elliptic curve cryptographic protocols generally require fast hardware and software implementations of the multiplication and inversion operations. On the other hand circuits for these operations for small fields may be implemented completely in hardware and/or using a table lookup approach.

The subject of this paper is multiplication algorithms in the binary extension fields $GF(2^k)$. There are essentially two categories of algorithms, based on the representation of field elements using polynomial basis or normal basis. In this paper, a new transformation method and a new multiplication algorithm for normal basis is introduced. First we will review the existing algorithms for both polynomial and normal bases, and then introduce the transformation method, which maps the elements of the field uniquely to the same set. This also slightly changes the definition of multiplication, as the product is computed in the transformed domain. The resulting algorithm is useful for applications where several normal basis multiplications are performed, as is the case for elliptic curve cryptography.

© Springer International Publishing Switzerland 2015
Ç. Koç et al. (Eds.): WAIFI 2014, LNCS 9061, pp. 61–80, 2015.
DOI: 10.1007/978-3-319-16277-5_4

2 Polynomial Basis Multiplication

In the polynomial basis a field element $a \in GF(2^k)$ is represented as a polynomial of degree less than or equal to $k-1$, written as $a(x) = \sum_{i=0}^{k-1} a_i x^i$ with coefficients $a_i \in \{0, 1\}$. The addition of two field elements a and b is accomplished by adding the coefficients a_i and b_i in $GF(2)$, which is the XOR of two binary values. On the other hand, the multiplication $c = a \cdot b$ is accomplished by first computing the degree $2k - 2$ product polynomial $c'(x) = a(x)b(x)$ and then reducing it modulo the irreducible polynomial $p(x)$ of degree k, in order to obtain the product $c(x)$ of degree at most $k - 1$:

$$c(x) = a(x)b(x) \bmod p(x) \ .$$

The multiplication of the individual coefficients a_i and b_i require 2-input AND gates, while the steps of the multiplication is accomplished by shift and XOR operations in software, or rewiring and XOR gates in hardware.

There are various polynomial basis multiplication algorithms; the work of Mastrovito is quite remarkable [15,16]. This was followed up in [9,20,26,28].

The properties of the irreducible polynomial $p(x)$ are also important, and not to be overlooked. In general, low Hamming weight irreducible polynomials [24], for example, trinomials and pentanomials are preferred. These yield many efficient algorithms [10,12,27]. A comprehensive list of polynomial basis multiplication algorithms can be found in [5].

A polynomial basis multiplication algorithm of interest is the Montgomery multiplication algorithm, proposed by Koç and Acar in [13]. This algorithm has three important properties that do not exist in the common algorithms found in the literature. The first property is that it works for general irreducible polynomials, not just special ones (such as trinomials, pentanomials, or all-one polynomials), making it more suitable for software implementations of cryptographic algorithms. The second property is that, it actually computes

$$\bar{c}(x) = \mathrm{MonPro}(\bar{a}(x), \bar{b}(x)) = \bar{a}(x)\bar{b}(x)x^{-k} \bmod p(x) \ , \tag{1}$$

instead of the usual $c(x) = a(x)b(x) \bmod p(x)$. This algorithm is actually the polynomial analogue of the Montgomery multiplication algorithm for integers [14,17]. In order to compute a field multiplication, the elements a and b are first *forward transformed* into the polynomial Montgomery domain

$$a \to \bar{a} : \bar{a}(x) = a(x)x^k \bmod p(x)$$
$$b \to \bar{b} : \bar{b}(x) = b(x)x^k \bmod p(x)$$

and then, the Montgomery product is computed

$$\begin{aligned}
\bar{c}(x) &= \bar{a}(x)\bar{b}(x)x^{-k} \bmod p(x) \\
&= a(x)x^k b(x)x^k x^{-k} \bmod p(x) \\
&= a(x)b(x)x^k \bmod p(x) \ ,
\end{aligned}$$

which is equal to $c(x)x^k$. When the result $\bar{c}$ needs to be transformed back to c, we use

$$\bar{c} \to c : c(x) = \bar{c}(x)x^{-k} \bmod p(x) \ .$$

Of course, in order to be useful, one should not be needing too many forward $c \to \bar{c}$ and backward $\bar{c} \to c$ transformations. This is never a problem for applications we are considering, such as elliptic curve cryptography, where tens of field multiplications are performed for each elliptic curve point addition and doubling operations, and hundreds of field multiplications are performed for elliptic curve point multiplication operations.

Finally, the third property of the Montgomery multiplication algorithm is that it is more suitable for software implementations for general irreducible polynomials, because the reduction proceeds word-by-word, due to the properties of the Montgomery multiplication for integers [14,17]. However, it can be argued that this is a moot point, since in most cases, we have low Hamming weight irreducible polynomials (trinomials and pentanomials) and there is no particular need for general irreducible polynomials [4].

Before closing this section we should also add that the Montgomery multiplication in $GF(2^k)$ is not the only transformative multiplication algorithm; there are also spectral methods [22,23], embedding techniques [25], and transformation of the normal basis elements into polynomials [8].

3   Normal Basis Multiplication

An element β of the field $GF(2^k)$ is called a normal element, if all 2^k elements of the field can be uniquely written as a linear sum of the powers of two powers of β as

$$a = \sum_{i=0}^{k-1} a_i \beta^{2^i} = a_0\beta + a_1\beta^2 + a_2\beta^4 + \cdots + a_{k-1}\beta^{2^{k-1}},$$

such that $a_i \in \{0,1\}$. Since the work of Kurt Wilhelm Sebastian Hensel in 1888, we know that there always exists a normal element for any prime p and integer k for the field $GF(p^k)$.

For the brevity of the notation, we will interchangeably use $\beta_i = \beta^{2^i}$, and furthermore, use 1 to represent the unity element in normal basis:

$$1 = \beta + \beta^2 + \beta^4 + \cdots + \beta^{2^{k-1}} = \beta_0 + \beta_1 + \beta_2 + \cdots + \beta_{k-1}.$$

The normal representation of an element in $GF(2^k)$ is particularly useful for squaring it. Notice that, since

$$\beta_k = \beta^{2^k} = \beta = \beta_0$$
$$\beta_i^2 = \beta^{2^i}\beta^{2^i} = \beta^{2^{i+1}} = \beta_{i+1} \ ,$$

given $a = (a_{k-1}a_{k-2}\cdots a_1a_0)$, we obtain a^2 as

$$
\begin{aligned}
a^2 &= (a_0\beta_0 + a_1\beta_1 + a_2\beta_2 + \cdots + a_{k-1}\beta_{k-1})^2 \\
&= a_0\beta_0^2 + a_1\beta_1^2 + a_2\beta_2^2 + \cdots + a_{k-1}\beta_{k-1}^2 \\
&= a_0\beta_1 + a_1\beta_2 + a_2\beta_3 + \cdots + a_{k-1}\beta_k \\
&= a_{k-1}\beta_0 + a_0\beta_1 + a_1\beta_2 + a_2\beta_3 + \cdots + a_{k-2}\beta_{k-1} \\
&= (a_{k-2}a_{k-3}\cdots a_1a_0a_{k-1}) \,,
\end{aligned}
$$

Therefore, the normal expression of a^2 is obtained by left-rotating the digits of the normal expression of a. The ease of squaring in normal basis is remarkable, but the multiplication is more complicated.

In the following we explain the steps of the normal basis multiplication, which will be used to develop a new transformation method and normal basis multiplication algorithm.

In order to describe the computational requirements of the normal basis multiplication, we follow the steps of the Massey-Omura algorithm [19,21], which gives the general outline for normal basis multiplication. Given the input operands a and b, the Massey-Omura multiplier first generates all partial products a_ib_j for $0 \leq i, j \leq k - 1$ using AND gates, and then sums these partial product terms using multi-operand adders (whose unit element is an XOR gate).

There are k^2 partial product terms a_ib_j, a computation that can be performed using k^2 2-input AND gates in a single AND gate delay. Decidedly this computation is optimal; k^2 is both upper and lower bound on the number of partial product terms, because all of them need to be computed.

However, in the computation of each of the product terms c_r for $0 \leq r \leq k-1$, we need only a subset of the k^2 partial product terms a_ib_j. According to the optimality argument [18] of the normal basis multiplication, the number of a_ib_j terms needed to compute any of c_r is at least $2k - 1$ for $GF(2^k)$. If there exists a normal basis for which the number of a_ib_j terms for computing any of c_r is exactly $2k - 1$ for $GF(2^k)$, then this normal basis is called *optimal*. It should be noted that optimal normal bases do not exist for every value of k in $GF(2^k)$, which is easily verified for small values of k using exhaustive search. All values of $k \leq 2000$ for which there is an optimal normal basis of $GF(2^k)$ are listed in [6].

Several constructions of optimal normal bases are given in [18], together with a conjecture that describes all finite binary field extensions which have an optimal normal basis. It was proven by Gao and Lenstra in [6,7] that the optimal normal basis constructions given in [18] are indeed all there is. These constructions are summarized in the theorem below:

Theorem 1. *An optimal normal basis for $GF(2^k)$ exist only in either of the following cases:*

1. *If $k + 1$ is prime and 2 is a primitive element in $\mathcal{Z}_{k+1}$, then each of the k nonunit $(k + 1)$th root of unity forms an optimal normal basis in $GF(2^k)$.*
2. *If $2k + 1$ is prime and*
   2a: Either, 2 is primitive in $\mathcal{Z}_{2k+1}$;

2b: Or, $2k+1 \equiv 3 \pmod 4$ and 2 generates quadratic residues in $\mathbb{Z}_{2k+1}$; then, $\beta = \gamma + \gamma^{-1}$ generates an optimal normal basis in $GF(2^k)$, where γ is a primitive $(2k+1)$th root of unity.

For historical reasons, the optimal normal bases that satisfy the first part of the above theorem are named Type 1, while the ones that follow from the second part are named Type 2 bases.

3.1   Optimal Normal Multiplication in $GF(2^2)$

The elements of $GF(2^2)$ expressed in polynomial basis are $\{0, 1, x, x+1\}$. There is only one irreducible polynomial of degree 2 over $GF(2)$, which is $p(x) = x^2+x+1$. Since $k+1 = 3$ is prime, and 2 is a primitive element in $\mathbb{Z}_3$ (because $2^1 = 2$ and $2^2 = 1$), the field $GF(2^2)$ has Type 1 optimal normal basis. The 2 nonunit 3rd roots of the unity in $GF(2^2)$ are the two optimal normal basis elements of $GF(2^2)$, and they are x and $x+1$, because $x^3 = (x+1)^3 = 1 \bmod p(x)$.

We illustrate the normal basis multiplication in $GF(2^2)$ using the optimal normal element $\beta = x$. Let the normal representations of two operands given as $a = a_0\beta + a_1\beta^2$ and $b = b_0\beta + b_1\beta^2$. The product c is equal to

$$c = a_0b_0\beta^2 + a_0b_1\beta^3 + a_1b_0\beta^3 + a_1b_1\beta^4.$$

This expansion contains the terms β^2, β^3 and β^4. First we need to obtain the normal representation of β^3. Since $\beta = x$, we have $\beta^2 = x^2 = x+1 \bmod p(x)$, and thus, $\beta^3 = x(x+1) = x^2 + x = 1 \bmod p(x)$. Furthermore, $\beta + \beta^2 = x+x+1 = 1 \bmod p(x)$, and thus, we have

$$\beta^0 = \beta + \beta^2 = 1,$$
$$\beta^1 = \beta = x,$$
$$\beta^2 = \beta^2 = x+1,$$
$$\beta^3 = \beta + \beta^2 = 1.$$

Substituting β^3 and β^4 with $\beta + \beta^2$ and β in the expansion of the product c, we obtain

$$c = a_0b_0\beta^2 + a_0b_1(\beta + \beta^2) + a_1b_0(\beta + \beta^2) + a_1b_1\beta$$
$$= (a_0b_1 + a_1b_0 + a_1b_1)\beta + (a_0b_0 + a_0b_1 + a_1b_0)\beta^2 \quad (2)$$

which gives the individual terms of the product c as

$$c_0 = a_0b_1 + a_1b_0 + a_1b_1$$
$$c_1 = a_0b_0 + a_0b_1 + a_1b_0 \quad (3)$$

We now define the $k \times k$ matrix $\boldsymbol{\lambda}$ such that $\lambda_{ij} = \beta^{2^i+2^j}$ for $0 \le i, j \le k-1$, which for $k = 2$ is given as

$$\boldsymbol{\lambda} = \begin{bmatrix} \beta^2 & \beta^3 \\ \beta^3 & \beta^4 \end{bmatrix} = \begin{bmatrix} \beta^2 & \beta + \beta^2 \\ \beta + \beta^2 & \beta \end{bmatrix} = \begin{bmatrix} 0 & 1 \\ 1 & 1 \end{bmatrix}\beta + \begin{bmatrix} 1 & 1 \\ 1 & 0 \end{bmatrix}\beta^2 = \boldsymbol{\lambda}^{(0)}\beta + \boldsymbol{\lambda}^{(1)}\beta^2 \quad (4)$$

We will explain some properties of λ in the following subsection, however, it suffices to say that the matrix λ has $k(2k-1) = 6$ entries (3 in each column or 3 in each row). Furthermore, when it is expanded into the powers of β, we obtain the 0-1 matrices $\lambda^{(0)}$ and $\lambda^{(1)}$, each of which has $2k - 1 = 3$ nonzero entries.

3.2   Optimal Normal Multiplication in GF(2^3)

We now illustrate the normal basis multiplication in GF(2^3), which has Type 2 optimal normal basis. We use the optimal normal element $\beta = x + 1$, and irreducible polynomial $p(x) = x^3 + x + 1$. In this section we will also describe certain properties of the λ matrices that are relevant to our proposed multiplication algorithm. Let a and b given as

$$a = a_0\beta + a_1\beta^2 + a_2\beta^4 \ ,$$
$$b = b_0\beta + b_1\beta^2 + b_2\beta^4 \ .$$

The product c would be

$$c = a_0b_0\beta^2 + a_0b_1\beta^3 + a_0b_2\beta^5 + a_1b_0\beta^3 + a_1b_1\beta^4 + a_1b_2\beta^6$$
$$+a_2b_0\beta^5 + a_2b_1\beta^6 + a_2b_2\beta^8 \ . \tag{5}$$

This expansion contains terms β^2, β^4, and β^8. Since $\beta^8 = \beta$, these are the powers of 2 powers of β, required for normal representation in GF(2^3). However, the above expansion of c also contains other powers: β^3, β^5, and β^6. All powers of β can be expressed in polynomial basis, reduced modulo the irreducible polynomial $p(x)$, generating a conversion table between the powers of β and the elements of the field represented in polynomial basis. Furthermore, once the polynomial representations of β, β^2 and β^4 are obtained, we can also obtain the normal representations of all elements. Table 1 contains the polynomial, the normal, and the powers of β representations of the field elements.

Table 1. The normal and the powers of $\beta = x + 1$ representations of elements in GF(2^3) with irreducible polynomial $p(x) = x^3 + x + 1$.

$\beta^4 + \beta^2 + \beta$	β^0
β	β^1
β^2	β^2
$\beta^4 + \beta$	β^3
β^4	β^4
$\beta^4 + \beta^2$	β^5
$\beta^2 + \beta$	β^6
$\beta^4 + \beta^2 + \beta$	β^7

Substituting the powers of β in the expansion of the product c in Eq. (5), we obtain

$$
\begin{aligned}
c &= a_0 b_0 \beta^2 + a_0 b_1 (\beta + \beta^4) + a_0 b_2 (\beta^2 + \beta^4) + a_1 b_0 (\beta + \beta^4) + a_1 b_1 \beta^4 + a_1 b_2 (\beta + \beta^2) \\
&\quad + a_2 b_0 (\beta^2 + \beta^4) + a_2 b_1 (\beta + \beta^2) + a_2 b_2 \beta \\
&= (a_0 b_1 + a_1 b_0 + a_2 b_2 + a_1 b_2 + a_2 b_1)\beta + (a_0 b_0 + a_0 b_2 + a_2 b_0 + a_1 b_2 + a_2 b_1)\beta^2 \\
&\quad + (a_0 b_1 + a_1 b_0 + a_0 b_2 + a_2 b_0 + a_1 b_1)\beta^4 .
\end{aligned}
$$

which gives the individual terms of the product c as

$$
\begin{aligned}
c_0 &= a_0 b_1 + a_1 b_0 + a_2 b_2 + a_1 b_2 + a_2 b_1, \\
c_1 &= a_0 b_0 + a_0 b_2 + a_2 b_0 + a_1 b_2 + a_2 b_1, \\
c_1 &= a_0 b_1 + a_1 b_0 + a_0 b_2 + a_2 b_0 + a_1 b_1 .
\end{aligned} \tag{6}
$$

The $k \times k$ matrix $\boldsymbol{\lambda}$ with entries $\lambda_{ij} = \beta^{2^i + 2^j}$ for $0 \le i, j \le k - 1$ is given as

$$
\boldsymbol{\lambda} = \begin{bmatrix} \beta^2 & \beta^3 & \beta^5 \\ \beta^3 & \beta^4 & \beta^6 \\ \beta^5 & \beta^6 & \beta^8 \end{bmatrix} = \begin{bmatrix} \beta^2 & \beta + \beta^4 & \beta^2 + \beta^4 \\ \beta + \beta^4 & \beta^4 & \beta + \beta^2 \\ \beta^2 + \beta^4 & \beta + \beta^2 & \beta \end{bmatrix} . \tag{7}
$$

The $\boldsymbol{\lambda}$ matrix contains all powers of β needed in the computation of c, as given in Eq. (5). It can also be expressed by separating the powers of β as

$$
\boldsymbol{\lambda} = \begin{bmatrix} 0 & 1 & 0 \\ 1 & 0 & 1 \\ 0 & 1 & 1 \end{bmatrix} \beta + \begin{bmatrix} 1 & 0 & 1 \\ 0 & 0 & 1 \\ 1 & 1 & 0 \end{bmatrix} \beta^2 + \begin{bmatrix} 0 & 1 & 1 \\ 1 & 1 & 0 \\ 1 & 0 & 0 \end{bmatrix} \beta^4 = \boldsymbol{\lambda}^{(0)}\beta + \boldsymbol{\lambda}^{(1)}\beta^2 + \boldsymbol{\lambda}^{(2)}\beta^4 , \tag{8}
$$

where, the 3×3 matrices $\boldsymbol{\lambda}^{(r)}$ for $r = 0, 1, 2$ have entries in $\{0, 1\}$. Since the product c in Eq. (5) can be written as

$$
\begin{aligned}
c &= \sum_{i=0}^{2} \sum_{j=0}^{2} a_i b_j \beta^{2^i + 2^j} = \sum_{i=0}^{2} \sum_{j=0}^{2} a_i b_j \lambda_{ij} \\
&= \sum_{i=0}^{2} \sum_{j=0}^{2} a_i b_j \lambda_{ij}^{(0)} \beta + \sum_{i=0}^{2} \sum_{j=0}^{2} a_i b_j \lambda_{ij}^{(1)} \beta^2 + \sum_{i=0}^{2} \sum_{j=0}^{2} a_i b_j \lambda_{ij}^{(2)} \beta^4
\end{aligned}
$$

By expressing c as $c = c_0 \beta + c_1 \beta^2 + c_2 \beta^4$, we can write the individual terms of the product c_r as

$$
c_r = \sum_{i=0}^{2} \sum_{j=0}^{2} a_i b_j \lambda_{ij}^{(r)} .
$$

The complexity of computing the terms c_r depends on the number of 1s in the matrices $\boldsymbol{\lambda}^{(r)}$ for $r = 0, 1, 2$. Furthermore, the matrices $\lambda_{ij}^{(r)}$ have the following properties:

$$
\lambda_{ij}^{(r)} = \lambda_{ji}^{(r)} \tag{9}
$$

$$
\lambda_{i+1, j+1}^{(r+1)} = \lambda_{i,j}^{(r)} \tag{10}
$$

The first property is due to the fact that $2^i + 2^j = 2^j + 2^i$, and thus,

$$\lambda_{ij} = \beta^{2^i+2^j} = \beta^{2^j+2^i} = \lambda_{ji} .$$

The second property follows from the fact that $2^{i+1} + 2^{j+1} = 2(2^i + 2^j)$, and thus

$$\beta^{2^{i+1}+2^{j+1}} = (\beta^2)^{2^i+2^j} ,$$

which implies

$$\lambda_{i+1,j+1}(\beta) = \lambda_{i,j}(\beta^2) .$$

Considering also the fact that $\beta^{2^k} = \beta$, we obtain Eq. (10). Note that all index arithmetic, i.e., increments such as $i+1$ and $j+1$ are considered mod 3 in GF(2^3) or mod k in GF(2^k).

The optimal basis theorem [7,18] teaches that if the normal element β is optimal, then each one of the matrices $\lambda_{ij}^{(r)}$ for $0 \leq r \leq k-1$ has $2k-1$ nonzero entries, as was the case for both GF(2^2) in Eq. (4) and GF(2^3) in Eq. (8), where we have $2 \cdot 2 - 1 = 3$ and $2 \cdot 3 - 1 = 5$ nonzero elements in each matrix. Equivalently, the matrix λ has $k(2k-1)$ individual terms such that each term is a power of 2 power of β, as shown in Eq. (4) for GF(2^2) and Eq. (7) for GF(2^3), which has 15 terms.

3.3   Complexity and Implementation

While there are several different ways of putting things together, the basic outline of a normal basis multiplier has 2 steps:

– Step 1: Compute $a_i b_j$ terms using k^2 2-input AND gates.
– Step 2: Sum the subset of the terms as implied by the nonzero entries of the $\lambda_{ij}^{(r)}$ matrix using $2k-2$ 2-input XOR gates for each c_r term.

Step 1 and Step 2 can be performed sequentially, partially parallel, or fully parallel. Since Step 1 is pretty obvious, that is, it computes k^2 different things, there is no need to dwell on it. Step 2, on the other hand, provides several different implementations and optimizations. For example, we can implement a single circuit consisting of $2k-2$ XOR gates (arranged either as a linear array or a binary tree) to compute c_0, and reuse the same circuit for computing c_r for $r = 1, 2, \ldots, k-1$, by only shifting the input operands a_i and b_j for $0 \leq i, j \leq k-1$. Figure 1 illustrates the construction.

We are intentionally ignoring some of the details of the circuit in Fig. 1, since there are various ways to arrange the circuit elements, for example, sequential, parallel, systolic, and pipelined circuits have been designed [1–3]. Our focus in this paper is not on how the individual steps of the optimal normal basis multiplications are performed or how individual circuit elements are arranged. Rather, we are interested in discovering whether there is another way to multiply two elements expressed in a normal basis defined by β in GF(2^k).

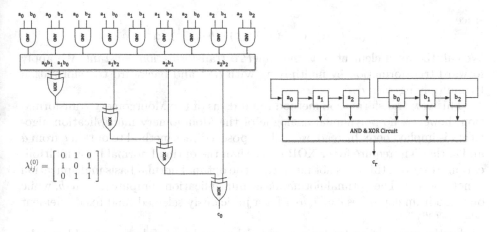

Fig. 1. Optimal normal basis multiplier construction for GF(2^3).

4   The Proposed Method

Let a and b be expressed in an optimal normal basis, using the normal element β in GF(2^k). The multiplication of a and b produces c, expressed as

$$c = \sum_{i=0}^{k-1}\sum_{j=0}^{k-1} a_i b_j \beta^{2^i+2^j} = \sum_{i=0}^{k-1}\sum_{j=0}^{k-1} a_i b_j \lambda_{ij} \,, \tag{11}$$

such that the $k \times k$ matrix $\boldsymbol{\lambda}$ has $k(2k-1)$ terms of type β^{2^i} for $i = 1, 2, \dots, $ k-1. Let α be a fixed element of GF(2^k), such that $\alpha \neq 0, 1$. We will also need α^{-1} which can be precomputed using the extended Euclidean algorithm or the Fermat's method, or the Itoh-Tsujii method [11], which is actually based on the Fermat's method.

We define a new multiplication function, which we denote as NewPro, that takes two elements $\bar{a}$ and $\bar{b}$ of the field GF(2^k), which are the forward transformations of a and b, as

$$a \rightarrow \bar{a}: \quad \bar{a} = a \cdot \alpha^{-1} \,, \tag{12}$$

$$b \rightarrow \bar{b}: \quad \bar{b} = b \cdot \alpha^{-1} \,. \tag{13}$$

The transformation requires the precomputed α^{-1} value. The operands a and b are now expressed in "bar" domain. The NewPro algorithm takes $\bar{a}$ and $\bar{b}$ as input and computes $\bar{c}$ as

$$\bar{c} = \text{NewPro}(\bar{a}, \bar{b}) = \bar{a} \cdot \bar{b} \cdot \alpha \,. \tag{14}$$

After the multiplication, the resulting $\bar{c}$ can be backward transformed to "nobar" domain using

$$\bar{c} \rightarrow c: \quad c = \bar{c} \cdot \alpha \,. \tag{15}$$

since

$$\bar{c} \cdot \alpha = (\bar{a} \cdot \bar{b} \cdot \alpha) \cdot \alpha = (a \cdot \alpha^{-1}) \cdot (b \cdot \alpha^{-1}) \cdot \alpha^2 = a \cdot b \ .$$

We call the fixed element α as the *NewPro transformation constant*. We apply forward transformation by multiplying with α^{-1} and backward transformation by multiplying with α.

This new transformation method reminds us of the Montgomery transformation, however, no polynomial analogue of the Montgomery multiplication algorithm is implied here. Instead, we will propose a direct method to obtain $\bar{c}$ from $\bar{a}$ and $\bar{b}$, that will require fewer XOR gates than the optimal normal basis multiplication. However, this does not mean that the optimal normal basis multiplication is not optimal. The optimal normal basis multiplication computes $c = a \cdot b$, while our algorithm computes $\bar{c} = \bar{a} \cdot \bar{b} \cdot \alpha$ for a judiciously selected (and fixed) element $\alpha \in \mathrm{GF}(2^k)$.

Furthermore, in order for our algorithm to be useful, one should not be needing too many forward $c \to \bar{c}$ and backward $\bar{c} \to c$ transformations. Again, this is not a problem for applications we are considering.

Before proceeding, we should also add that both forward and backward transformations are trivially performed using the NewPro algorithm:

$$
\begin{aligned}
a \to \bar{a} &: \mathrm{NewPro}(a, \alpha^{-2}) = a \cdot \alpha^{-2} \cdot \alpha = \bar{a} \ , \\
\bar{a} \to a &: \mathrm{NewPro}(\bar{a}, 1) = (a \cdot \alpha^{-1}) \cdot 1 \cdot \alpha = a.
\end{aligned}
\tag{16}
$$

For these computations, we need α^{-2}, which is easily obtained from the normal representation of α^{-1} by a left rotation of the digits. We also need the normal representation of the unity element, which is given as $(11 \cdots 1) = \sum_{i=0}^{k-1} \beta^{2^i}$ for any normal element β.

Also, if two elements are expressed in the bar domain, then their additions produce the output in the bar domain, that is

$$\bar{a} + \bar{b} = a \cdot \alpha^{-1} + b \cdot \alpha^{-1} = (a + b) \cdot \alpha^{-1} = c \cdot \alpha^{-1} = \bar{c} \ .$$

Finally we should remark that, similar to (11), the NewPro function for computing $\bar{c} = \bar{a} \cdot \bar{b} \cdot \alpha$ can be expanded as

$$\bar{c} = \left(\sum_{i=0}^{k-1} \sum_{j=0}^{k-1} \bar{a}_i \bar{b}_j \boldsymbol{\lambda}_{ij} \right) \cdot \alpha = \sum_{i=0}^{k-1} \sum_{j=0}^{k-1} \bar{a}_i \bar{b}_j (\alpha \boldsymbol{\lambda}_{ij}) \ .$$

This is the usual normal basis multiplication, however, the matrix involved is $\alpha \boldsymbol{\lambda}$, instead of just $\boldsymbol{\lambda}$, for a fixed element α of the field $\mathrm{GF}(2^k)$. The matrix $\boldsymbol{\lambda}$ has $k(2k-1)$ terms of type β^{2^i}, which determines the number of 2-input XOR gates as $2k - 2$ for computing each component of c_r for $0 \le r \le k - 1$.

In order to have a reduced complexity (in terms of the XOR gates) normal basis multiplication, we need to show that there exists a special element α of $\mathrm{GF}(2^k)$ for which $\alpha \boldsymbol{\lambda}$ has fewer than $k(2k-1)$ terms of type β^{2^i}. We will show the construction of α and the analyses for the fields $\mathrm{GF}(2^2)$, $\mathrm{GF}(2^3)$, and $\mathrm{GF}(2^4)$ below, and then describe the general cases.

4.1   Complexity of NewPro Multiplication in $GF(2^2)$

The proposed NewPro algorithm requires the existence of a special element α of $GF(2^k)$ such that the matrix $\alpha\lambda$ has fewer than $k(2k-1)$ terms of type β^{2^i}. Since $GF(2^2)$ has only two suitable elements β and β^2, we can easily try each one to see if the number of terms in the matrix $\alpha\lambda$ is less than 6. First we consider $\alpha = \beta$:

$$\alpha\lambda = \beta\lambda = \beta \begin{bmatrix} \beta^2 & \beta + \beta^2 \\ \beta + \beta^2 & \beta \end{bmatrix} = \begin{bmatrix} \beta^3 & \beta^2 + \beta^3 \\ \beta^2 + \beta^3 & \beta^2 \end{bmatrix} = \begin{bmatrix} \beta + \beta^2 & \beta \\ \beta & \beta^2 \end{bmatrix}$$

Indeed the resulting $\alpha\lambda$ has 5 terms, instead of 6. This gives the $\alpha\lambda^{(r)}$ matrices as

$$\alpha\lambda = \beta\lambda = \begin{bmatrix} 1 & 1 \\ 1 & 0 \end{bmatrix} \beta + \begin{bmatrix} 1 & 0 \\ 0 & 1 \end{bmatrix} \beta^2$$

We obtain the individual components of the product $\bar{c}$ as

$$\bar{c}_0 = \bar{a}_0 \bar{b}_0 + \bar{a}_0 \bar{b}_1 + \bar{a}_1 \bar{b}_0 \ ,$$
$$\bar{c}_0 = \bar{a}_0 \bar{b}_0 + \bar{a}_1 \bar{b}_0 \ .$$

Therefore we showed that the NewPro multiplication $\bar{c} = \text{NewPro}(\bar{a}, \bar{b}) = \bar{a}\bar{b}\alpha$ requires only 3 2-input XOR gates using the above formulae with the selection of $\alpha = \beta$, instead of 4 2-input XOR gates required by the normal product computation $c = ab$, as given by the formulae in Eq. (3).

It turns out that $\alpha = \beta^2$ also reduces the complexity:

$$\alpha\lambda = \beta^2\lambda = \beta^2 \begin{bmatrix} \beta^2 & \beta + \beta^2 \\ \beta + \beta^2 & \beta \end{bmatrix} = \begin{bmatrix} \beta^4 & \beta^3 + \beta^4 \\ \beta^3 + \beta^4 & \beta^3 \end{bmatrix} = \begin{bmatrix} \beta & \beta^2 \\ \beta^2 & \beta + \beta^2 \end{bmatrix}$$

This gives the $\alpha\lambda^{(r)}$ matrices as

$$\alpha\lambda = \begin{bmatrix} 1 & 0 \\ 0 & 1 \end{bmatrix} \beta + \begin{bmatrix} 0 & 1 \\ 1 & 1 \end{bmatrix} \beta^2$$

We obtain the individual components of the product $\bar{c}$ as

$$\bar{c}_0 = \bar{a}_0 \bar{b}_0 + \bar{a}_1 \bar{b}_1 \ ,$$
$$\bar{c}_0 = \bar{a}_0 \bar{b}_1 + \bar{a}_1 \bar{b}_0 + \bar{a}_1 \bar{b}_1 \ .$$

The NewPro multiplication $\bar{c} = \text{NewPro}(\bar{a}, \bar{b}) = \bar{a}\bar{b}\alpha$ requires only 3 2-input XOR gates using the above formulae with the selection of $\alpha = \beta^2$, instead of 4 2-input XOR gates required by the regular normal basis multiplication $c = ab$, as given by the formulae in Eq. (3).

4.2   Complexity of NewPro Multiplication in $GF(2^3)$

The proposed NewPro algorithm requires the existence of a special element α of $GF(2^k)$ such that the matrix $\alpha\lambda$ has fewer than $k(2k-1)$ terms of type β^{2^i}.

Table 2. The number of terms in the matrix $\alpha\lambda$ for $GF(2^3)$.

α	Terms
β	17
β^2	17
β^4	17
$\beta + \beta^2$	14
$\beta + \beta^4$	14
$\beta^2 + \beta^4$	14

For a small field such as $GF(2^3)$, we can try all possible candidates for α. Since our construction excludes α as 0 or 1, we need to try only 6 different α values in $GF(2^3)$. We have performed this search using a simple Mathematica code, and obtained the number of elements in the $\alpha\lambda$ matrix for each value of α, as shown in Table 2. Note that without the transformation, the matrix λ has $k(2k-1) = 15$ terms.

The search shows that for $\alpha = \beta + \beta^2$, $\alpha = \beta + \beta^4$, and $\alpha = \beta^2 + \beta^4$ values the matrix $\alpha\lambda$ has only 14 terms, which is 1 fewer than 15, the optimal value for the matrix λ.

We show how to obtain the $\alpha\lambda$ matrix for $\alpha = \beta + \beta^2$. We first multiply $(\beta + \beta^2)$ with every term of the matrix λ, and then substitute the powers of β which are not powers of 2, using the normal representations of all elements given in Sect. 3.1.

$$\alpha\lambda = (\beta + \beta^2) \begin{bmatrix} \beta^2 & \beta + \beta^4 & \beta^2 + \beta^4 \\ \beta + \beta^4 & \beta^4 & \beta + \beta^2 \\ \beta^2 + \beta^4 & \beta + \beta^2 & \beta \end{bmatrix} = \begin{bmatrix} \beta & \beta^2 & \beta^4 \\ \beta^2 & \beta + \beta^4 & \beta^2 + \beta^4 \\ \beta^4 & \beta^2 + \beta^4 & \beta + \beta^2 + \beta^4 \end{bmatrix}.$$

As is observed, this matrix has 14 terms. We can expand this matrix in terms of $\lambda^{(r)}$ matrices and powers of β, to obtain

$$\alpha\lambda = \begin{bmatrix} 1 & 0 & 0 \\ 0 & 1 & 0 \\ 0 & 0 & 1 \end{bmatrix} \beta + \begin{bmatrix} 0 & 1 & 0 \\ 1 & 0 & 1 \\ 0 & 1 & 1 \end{bmatrix} \beta^2 + \begin{bmatrix} 0 & 0 & 1 \\ 0 & 1 & 1 \\ 1 & 1 & 1 \end{bmatrix} \beta^4 .$$

Since we destroyed the shift property (Eq. 10) of $\lambda^{(r)}$ matrices by multiplying α with λ, these matrices no longer have the same number of 1s, however, the total number of 1s is $3 + 5 + 6 = 14$, instead of $5 + 5 + 5 = 15$. Finally, we obtain the individual components of the $\bar{c}$ vector as

$$\bar{c}_0 = \bar{a}_0\bar{b}_0 + \bar{a}_1\bar{b}_1 + \bar{a}_2\bar{b}_2 ,$$
$$\bar{c}_1 = \bar{a}_0\bar{b}_1 + \bar{a}_1\bar{b}_0 + \bar{a}_1\bar{b}_2 + \bar{a}_2\bar{b}_1 + \bar{a}_2\bar{b}_2 ,$$
$$\bar{c}_2 = \bar{a}_0\bar{b}_2 + \bar{a}_1\bar{b}_1 + \bar{a}_1\bar{b}_2 + \bar{a}_2\bar{b}_0 + \bar{a}_2\bar{b}_1 + \bar{a}_2\bar{b}_2 .$$

which requires 11 2-input XOR gates, instead of 12. The other two α values also produce the λ matrices with exactly 14 terms, and the associated formulae for the components $\bar{c}_r$ are easily obtained.

4.3   Complexity of NewPro Multiplication in $GF(2^4)$

Given the irreducible polynomial $p(x) = x^4 + x + 1$ generating the field $GF(2^4)$, we obtain an optimal normal element $\beta = x^3$ using the construction in Theorem 1. This field has Type 1 optimal normal basis since $k + 1 = 5$ is prime, and 2 is primitive in $\mathcal{Z}_5$ (Table 3).

Table 3. The normal and the powers of $\beta = x^3$ representations of elements in $GF(2^4)$ with irreducible polynomial $p(x) = x^4 + x + 1$.

$\beta^8 + \beta^4 + \beta^2 + \beta$	β^0	β^8	β^8
β	β^1	β^9	β^4
β^2	β^2	β^{10}	$\beta^8 + \beta^4 + \beta^2 + \beta$
β^8	β^3	β^{11}	β
β^4	β^4	β^{12}	β^2
$\beta^8 + \beta^4 + \beta^2 + \beta$	β^5	β^{13}	β^8
β	β^6	β^{14}	β^4
β^2	β^7	β^{15}	$\beta^8 + \beta^4 + \beta^2 + \beta$

The 4×4 λ matrix is obtained as

$$\begin{bmatrix} \beta^2 & \beta^3 & \beta^5 & \beta^9 \\ \beta^3 & \beta^4 & \beta^6 & \beta^{10} \\ \beta^5 & \beta^6 & \beta^8 & \beta^{12} \\ \beta^9 & \beta^{10} & \beta^{12} & \beta^{16} \end{bmatrix} = \begin{bmatrix} \beta^2 & \beta^8 & \beta^8 + \beta^4 + \beta^2 + \beta & \beta^4 \\ \beta^8 & \beta^4 & \beta & \beta^8 + \beta^4 + \beta^2 + \beta \\ \beta^8 + \beta^4 + \beta^2 + \beta & \beta & \beta^8 & \beta^2 \\ \beta^4 & \beta^8 + \beta^4 + \beta^2 + \beta & \beta^2 & \beta \end{bmatrix}$$

The number of terms in the λ matrix for an optimal basis $\beta \in GF(2^4)$ is equal to $k(2k - 1) = 28$. The matrix λ can be expanded as in terms of $\lambda^{(r)}$ matrices and powers of 2 powers of β as

$$\lambda = \begin{bmatrix} 0&0&1&0 \\ 0&0&1&1 \\ 1&1&0&0 \\ 0&1&0&1 \end{bmatrix} \beta + \begin{bmatrix} 1&0&1&0 \\ 0&0&0&1 \\ 1&0&0&1 \\ 0&1&1&0 \end{bmatrix} \beta^2 + \begin{bmatrix} 0&0&1&1 \\ 0&1&0&1 \\ 1&0&0&0 \\ 1&1&0&0 \end{bmatrix} \beta^4 + \begin{bmatrix} 0&1&1&0 \\ 1&0&0&1 \\ 1&0&1&0 \\ 0&1&0&0 \end{bmatrix} \beta^8 .$$

The total number of 1s in these matrices is also 28, each of which has 7 1s. Therefore, we need 6 2-input XOR gates to computes each one of the c_r terms for $r = 0, 1, 2, 3$, which totals to 24 XOR gates. Similar to the case $GF(2^3)$, we performed an exhaustive search over the set $GF(2^4)$ and obtained the list of α values, and the minimum number of terms in the matrix $\alpha\lambda$, as summarized in Table 4.

Table 4. The minimum number of terms in the matrix $\alpha\lambda$ for $GF(2^4)$.

α	Terms
β	25
β^2	25
β^4	25
β^8	25

It turns out that there are only 4 α values, which minimize the number of terms in the matrix $\alpha\lambda$; the minimum value is found as 25. We obtain the matrix $\alpha\lambda$ for $\alpha = \beta$, by multiplying every element of the matrix by α. The elements of the matrix which contains the powers of β which are not powers of 2 are then replaced with their normal expansions.

$$
\alpha\lambda = \begin{bmatrix}
\beta^8 & \beta^4 & \beta & \beta+\beta^2+\beta^4+\beta^8 \\
\beta^4 & \beta+\beta^2+\beta^4+\beta^8 & \beta^2 & \beta \\
\beta & \beta^2 & \beta^4 & \beta^8 \\
\beta+\beta^2+\beta^4+\beta^8 & \beta & \beta^8 & \beta^2
\end{bmatrix}.
$$

The matrix $\alpha\lambda$ has exactly 25 terms. It can be expanded in terms of $\lambda^{(r)}$ matrices and powers of 2 powers of β as

$$
\alpha\lambda = \begin{bmatrix} 0&0&1&1 \\ 0&1&0&1 \\ 1&0&0&0 \\ 1&1&0&0 \end{bmatrix} \beta + \begin{bmatrix} 0&0&0&1 \\ 0&1&1&0 \\ 0&1&0&0 \\ 1&0&0&1 \end{bmatrix} \beta^2 + \begin{bmatrix} 0&1&0&1 \\ 1&1&0&0 \\ 0&0&1&0 \\ 1&0&0&0 \end{bmatrix} \beta^4 + \begin{bmatrix} 1&0&0&1 \\ 0&1&0&0 \\ 0&0&0&1 \\ 1&0&1&0 \end{bmatrix} \beta^8.
$$

The total number of 1 s in these matrices is also equal to 25. The number of 2-input XOR gates to compute $\bar{c}_r$ terms for $r = 0, 1, 2, 3$ is found as $6 + 5 + 5 + 5 = 21$.

5   The General Case for $GF(2^k)$

The NewPro transformation and the multiplication algorithms require the existence of an element α of $GF(2^k)$, that minimizes the number of terms in the $\alpha\lambda$ matrix. We gave detailed analyses of the NewPro multiplication for the fields $GF(2^2)$, $GF(2^3)$ and $GF(2^4)$ together with all special α values. It turns out that the number of terms in the $\alpha\lambda$ matrix are equal to 5, 14 and 25 for $GF(2^2)$, $GF(2^3)$ and $GF(2^4)$ respectively, while the original λ matrices have 6, 15 and 28 terms.

However, we need a more detailed analysis of the proposed NewPro algorithm, specifically, we identify the following types of problems to study:

1. Due to the optimal normal basis theorem, we know that λ has $k(2k-1)$ terms for $GF(2^k)$. What is the exact number of terms in the $\alpha\lambda$ matrix for $GF(2^k)$ for different values of k and different (Type 1 and 2) bases?

2. Does there exist an α for any k such that number of terms in $\alpha\lambda$ is less than $k(2k-1)$?
3. Is there a constructive or non-exhaustive method for finding α that reduces the number of terms to fewer than $k(2k-1)$?

The answers to these questions for Type 2 optimal normal bases seem to be negative for $k > 3$. For such normal bases, no α can bring down the number of terms in $\alpha\lambda$ to a quantity below $k(2k-1)$. The case for $k = 3$, in which we were able to reduce the number of terms from 15 to 14, seems to be special. We do not have a complete proof of these assertions on Type 2.

However, we settle the above questions for Type 1 optimal normal bases in the following fashion.

6   Optimality for the Type 1 Case

Let us assume that $GF(2^k)$ has a Type 1 optimal normal basis; this implies that $k+1$ is prime and 2 is primitive in $\mathbb{Z}_{k+1}$. Moreover, the optimal normal element β is a primitive $(k+1)$st root of 1 in $GF(2^k)$. For the brevity of the notation, we write $k = 2m$ and use $\mathcal{B}$ to represent the basis set $\mathcal{B} = \{\beta_0, \beta_1, \ldots, \beta_{k-1}\}$. As before, the $k \times k$ matrix λ is defined as

$$\lambda_{ij} = \beta^{2^i + 2^j} = \beta_i \beta_j$$

for $0 \leq i, j \leq k-1$. For example, for $k = 4$, we have

$$\lambda = \begin{bmatrix} \beta_1 & \beta_3 & 1 & \beta_2 \\ \beta_3 & \beta_2 & \beta_0 & 1 \\ 1 & \beta_0 & \beta_3 & \beta_1 \\ \beta_2 & 1 & \beta_1 & \beta_0 \end{bmatrix}.$$

Lemma 1. *The elements in the entries* $(0, m), (1, m+1), \ldots, (k-1, m+k-1)$ *of* λ, *where the indices are computed mod* k, *are all* 1*s.*

Proof. What we need to show is $\beta_i \beta_{m+i} = 1$ for $0 \leq i \leq k-1$, where the indices are computed mod k. Let $\theta_i = \beta_i \beta_{m+i}$, and put $\theta = \theta_0 = \beta_0 \beta_m$. Then

$$\theta_i = \beta^{2^i + 2^{m+i}} = \left(\beta^{2^m + 1}\right)^{2^i} = \theta^{2^i}.$$

Therefore it suffices to show that $\theta = 1$. Calculating,

$$\theta^{2^m} = \beta^{2^m} \beta^{2^{2m}} = \beta_m \beta_0 = \theta,$$

so that $\theta^{2^m - 1} = 1$. On the other hand, θ is a power of β and $\beta^{2m+1} = 1$, so $\theta^{2m+1} = 1$. Therefore the order of θ divides $d = \gcd(2^m - 1, 2m + 1)$. Since $p = 2m + 1$ is prime, d is either 1 or p. But $2^m - 1 = 2^{\frac{p-1}{2}} - 1$ and this cannot be divisible by p for otherwise 2 is a quadratic residue modulo p and so cannot be primitive. Therefore, $d = 1$ and $\theta = 1$. $\square$

The entry $\beta_i\beta_{m+i} = \beta_0 + \beta_1 + \cdots + \beta_{k-1}$ contributes k to the sum of the number of basis vectors appearing in row i. Since the basis is optimal, the total number of these is $2k-1$. Each of the remaining $k-1$ entries is a single β_j. By optimality, the elements in row i excluding the unit in column $m+i$ is a permutation of $\beta_0, \ldots, \beta_{i-1}, \beta_{i+1}, \ldots, \beta_{k-1}$. We record the fact that the elements in row r of λ is a permutation of the elements $\mathcal{B} - \{\beta_r\}$.

Lemma 2. *For an optimal normal basis of Type 1 with $k+1 = 2m+1$, generated by $\beta = \beta_0$, the row r for $0 \le r \le k-1$ of λ is a permutation of $\mathcal{B} - \{\beta_r\}$ with 1 appearing in the column index $m+r$ modulo k. Therefore $\beta_r \cdot \{\beta_0, \beta_1, \ldots, \beta_{k-1}\} = \mathcal{B} - \{\beta_r\}$.*

Next, we consider the matrix $\beta_r \lambda$.

Lemma 3. *Row $m+r$ of $\beta_r \lambda$ is a permutation of $\mathcal{B}$. Each of the other rows is a permutation of $\mathcal{B}$ minus some basis element.*

Proof. We will give the proof for $r = 0$. Note that every row in λ has a 1 (the entries in positions $(0, m), (1, m+1), \ldots, (k-1, m+k-1)$ are 1s), so in $\beta\lambda$, β_0 appears in each row. By Lemma 2, the rth row of λ is a permutation of $\mathcal{B} - \{\beta_r\}$. Therefore the rth row of $\beta\lambda$ is

1. a permutation of $\mathcal{B} - \{1\}$ for $r = m$,
2. a permutation of $\mathcal{B} - \{\beta_0\beta_r\}$ for $r \ne m$ (note: $\beta_0\beta_r = \beta_j$ for some j in this case). $\qquad\square$

Therefore, we conclude that the total number of basis vectors appearing in the matrix is

$$k + (k-1)(2k-1) = k(2k-1) - (k-1) \, ,$$

which is $k-1$ fewer than that of the multiplication matrix λ. We state the following theorem for the Type 1 case, but omit its proof.

Theorem 2. *Suppose α has t nonzero coefficients in its normal basis expansion. Then the number of terms in the matrix $\alpha\lambda$ is*

$$k(2k-1) + (k-t)\left(t(2k-2) - (2k-1)\right) \, .$$

In particular $\alpha = \beta_i$ for $0 \le i \le k-1$ are the only αs with the property that $\alpha\lambda$ matrix has fewer than $k(2k-1)$, i.e., exactly $k(2k-1) - (k-1)$ basis vectors.

7   Further Work

By slightly changing the definition of the multiplication operation, we introduced a new normal basis multiplication algorithm which requires fewer XORs than the optimal normal multiplication algorithm. We proved for the Type 1 case that the number of terms in the $\alpha\lambda$ is $k-1$ fewer than that of λ matrix for $\alpha = \beta_i$ for some $0 \le i \le k-1$. Appendix A gives the λ and $\alpha\lambda$ matrices for $GF(2^k)$

for $k = 2, 4, 10, 12$, which are the first 4 fields that has Type 1 optimal normal bases.

Moreover, experimentation shows that the field $GF(2^3)$ and Type 2 optimal normal basis matrix $\alpha\lambda$ has 14 terms for $\alpha = \beta + \beta^2$, $\alpha = \beta + \beta^4$, and $\alpha = \beta^2 + \beta^4$ instead of 15 terms, which is clearly unlike the Type 1 case, i.e., it is not $k - 1 = 2$ fewer than $k(2k - 1) = 15$. However the case of $GF(2^3)$ seems to be an anomaly, and it appears that for a Type 2 optimal normal basis $GF(2^k)$ for $k > 3$, there is no α value which gives smaller than $k(2k - 1)$ terms in $\alpha\lambda$. However, we do not have a complete proof at this time.

It is also possible to view $\alpha\lambda$ as matrix multiplication by αI, so one can study the suitability of transformations defined by premultiplying λ by other kinds of matrices.

Appendix A: The Type 1 λ and $\alpha\lambda$ Matrices

The λ and $\alpha\lambda$ Matrices for $GF(2^2)$

The irreducible polynomial is $x^2 + x + 1$. The optimal normal element is $\beta = x$. The total count of 1 s in λ and $\alpha\lambda$ matrices are $k(2k - 1) = 6$ and $k(2k - 1) - (k - 1) = 5$, respectively.

$$\lambda = \begin{bmatrix} \beta_1 & \beta_0 \\ \beta_0 & \beta_1 \end{bmatrix} \quad , \quad \beta_0\lambda = \begin{bmatrix} 1 & \beta_0 \\ \beta_0 & \beta_1 \end{bmatrix} \quad , \quad \beta_1\lambda = \begin{bmatrix} \beta_0 & \beta_1 \\ \beta_1 & 1 \end{bmatrix}$$

The λ and $\alpha\lambda$ Matrices for $GF(2^4)$

The irreducible polynomial is $x^4 + x^3 + 1$. The optimal normal element is $\beta = x + 1$. The total count of 1 s in λ and $\alpha\lambda$ matrices are $k(2k - 1) = 28$ and $k(2k - 1) - (k - 1) = 25$, respectively.

$$\lambda = \begin{bmatrix} \beta_1 & \beta_3 & 1 & \beta_2 \\ \beta_3 & \beta_2 & \beta_0 & 1 \\ 1 & \beta_0 & \beta_3 & \beta_1 \\ \beta_2 & 1 & \beta_1 & \beta_0 \end{bmatrix} \quad , \quad \beta_0\lambda = \begin{bmatrix} \beta_3 & \beta_2 & \beta_0 & 1 \\ \beta_2 & 1 & \beta_1 & \beta_0 \\ \beta_0 & \beta_1 & \beta_2 & \beta_3 \\ 1 & \beta_0 & \beta_3 & \beta_1 \end{bmatrix} \quad , \quad \beta_1\lambda = \begin{bmatrix} \beta_2 & 1 & \beta_1 & \beta_0 \\ 1 & \beta_0 & \beta_3 & \beta_1 \\ \beta_1 & \beta_3 & 1 & \beta_2 \\ \beta_0 & \beta_1 & \beta_2 & \beta_3 \end{bmatrix}$$

$$\beta_2\lambda = \begin{bmatrix} \beta_0 & \beta_1 & \beta_2 & \beta_3 \\ \beta_1 & \beta_3 & 1 & \beta_2 \\ \beta_2 & 1 & \beta_1 & \beta_0 \\ \beta_3 & \beta_2 & \beta_0 & 1 \end{bmatrix} \quad , \quad \beta_3\lambda = \begin{bmatrix} 1 & \beta_0 & \beta_3 & \beta_1 \\ \beta_0 & \beta_1 & \beta_2 & \beta_3 \\ \beta_3 & \beta_2 & \beta_0 & 1 \\ \beta_1 & \beta_3 & 1 & \beta_2 \end{bmatrix}$$

The λ and $\alpha\lambda$ Matrices for $GF(2^{10})$

The irreducible polynomial is $x^{10} + x^7 + 1$. The optimal normal element is $\beta = x^6 + x^3 + x^2 + x$. The total count of 1 s in λ and $\alpha\lambda$ matrices are $k(2k-1) = 190$ and $k(2k-1) - (k-1) = 181$, respectively. Below we give λ and only $\beta_0\lambda$ matrix.

$$
\lambda =
\begin{bmatrix}
\beta_1 & \beta_8 & \beta_4 & \beta_6 & \beta_9 & 1 & \beta_5 & \beta_3 & \beta_2 & \beta_7 \\
\beta_8 & \beta_2 & \beta_9 & \beta_5 & \beta_7 & \beta_0 & 1 & \beta_6 & \beta_4 & \beta_3 \\
\beta_4 & \beta_9 & \beta_3 & \beta_0 & \beta_6 & \beta_8 & \beta_1 & 1 & \beta_7 & \beta_5 \\
\beta_6 & \beta_5 & \beta_0 & \beta_4 & \beta_1 & \beta_7 & \beta_9 & \beta_2 & 1 & \beta_8 \\
\beta_9 & \beta_7 & \beta_6 & \beta_1 & \beta_5 & \beta_2 & \beta_8 & \beta_0 & \beta_3 & 1 \\
1 & \beta_0 & \beta_8 & \beta_7 & \beta_2 & \beta_6 & \beta_3 & \beta_9 & \beta_1 & \beta_4 \\
\beta_5 & 1 & \beta_1 & \beta_9 & \beta_8 & \beta_3 & \beta_7 & \beta_4 & \beta_0 & \beta_2 \\
\beta_3 & \beta_6 & 1 & \beta_2 & \beta_0 & \beta_9 & \beta_4 & \beta_8 & \beta_5 & \beta_1 \\
\beta_2 & \beta_4 & \beta_7 & 1 & \beta_3 & \beta_1 & \beta_0 & \beta_5 & \beta_9 & \beta_6 \\
\beta_7 & \beta_3 & \beta_5 & \beta_8 & 1 & \beta_4 & \beta_2 & \beta_1 & \beta_6 & \beta_0
\end{bmatrix}
,\quad
\beta_0\lambda =
\begin{bmatrix}
\beta_8 & \beta_2 & \beta_9 & \beta_5 & \beta_7 & \beta_0 & 1 & \beta_6 & \beta_4 & \beta_3 \\
\beta_2 & \beta_4 & \beta_7 & 1 & \beta_3 & \beta_1 & \beta_0 & \beta_5 & \beta_9 & \beta_6 \\
\beta_9 & \beta_7 & \beta_6 & \beta_1 & \beta_5 & \beta_2 & \beta_8 & \beta_0 & \beta_3 & 1 \\
\beta_5 & 1 & \beta_1 & \beta_9 & \beta_8 & \beta_3 & \beta_7 & \beta_4 & \beta_0 & \beta_2 \\
\beta_7 & \beta_3 & \beta_5 & \beta_8 & 1 & \beta_4 & \beta_2 & \beta_1 & \beta_6 & \beta_0 \\
\beta_0 & \beta_1 & \beta_2 & \beta_3 & \beta_4 & \beta_5 & \beta_6 & \beta_7 & \beta_8 & \beta_9 \\
1 & \beta_0 & \beta_8 & \beta_7 & \beta_2 & \beta_6 & \beta_3 & \beta_9 & \beta_1 & \beta_4 \\
\beta_6 & \beta_5 & \beta_0 & \beta_4 & \beta_1 & \beta_7 & \beta_9 & \beta_2 & 1 & \beta_8 \\
\beta_4 & \beta_9 & \beta_3 & \beta_0 & \beta_6 & \beta_8 & \beta_1 & 1 & \beta_7 & \beta_5 \\
\beta_3 & \beta_6 & 1 & \beta_2 & \beta_0 & \beta_9 & \beta_4 & \beta_8 & \beta_5 & \beta_1
\end{bmatrix}
$$

The λ and $\alpha\lambda$ Matrices for $GF(2^{12})$

The irreducible polynomial is $x^{12} + x^{10} + x^2 + x + 1$. The optimal normal element is $\beta = x^{11} + x^7 + x^3 + x^2 + x$. The total count of 1 s in λ and $\alpha\lambda$ matrices are $k(2k-1) = 276$ and $k(2k-1) - (k-1) = 265$, respectively. Below we give λ and only $\beta_0\lambda$ matrix.

$$
\lambda =
\begin{bmatrix}
\beta_1 & \beta_4 & \beta_9 & \beta_8 & \beta_2 & \beta_{11} & 1 & \beta_6 & \beta_{10} & \beta_5 & \beta_7 & \beta_3 \\
\beta_4 & \beta_2 & \beta_5 & \beta_{10} & \beta_9 & \beta_3 & \beta_0 & 1 & \beta_7 & \beta_{11} & \beta_6 & \beta_8 \\
\beta_9 & \beta_5 & \beta_3 & \beta_6 & \beta_{11} & \beta_{10} & \beta_4 & \beta_1 & 1 & \beta_8 & \beta_0 & \beta_7 \\
\beta_8 & \beta_{10} & \beta_6 & \beta_4 & \beta_7 & \beta_0 & \beta_{11} & \beta_5 & \beta_2 & 1 & \beta_9 & \beta_1 \\
\beta_2 & \beta_9 & \beta_{11} & \beta_7 & \beta_5 & \beta_8 & \beta_1 & \beta_0 & \beta_6 & \beta_3 & 1 & \beta_{10} \\
\beta_{11} & \beta_3 & \beta_{10} & \beta_0 & \beta_8 & \beta_6 & \beta_9 & \beta_2 & \beta_1 & \beta_7 & \beta_4 & 1 \\
1 & \beta_0 & \beta_4 & \beta_{11} & \beta_1 & \beta_9 & \beta_7 & \beta_{10} & \beta_3 & \beta_2 & \beta_8 & \beta_5 \\
\beta_6 & 1 & \beta_1 & \beta_5 & \beta_0 & \beta_2 & \beta_{10} & \beta_8 & \beta_{11} & \beta_4 & \beta_3 & \beta_9 \\
\beta_{10} & \beta_7 & 1 & \beta_2 & \beta_6 & \beta_1 & \beta_3 & \beta_{11} & \beta_9 & \beta_0 & \beta_5 & \beta_4 \\
\beta_5 & \beta_{11} & \beta_8 & 1 & \beta_3 & \beta_7 & \beta_2 & \beta_4 & \beta_0 & \beta_{10} & \beta_1 & \beta_6 \\
\beta_7 & \beta_6 & \beta_0 & \beta_9 & 1 & \beta_4 & \beta_8 & \beta_3 & \beta_5 & \beta_1 & \beta_{11} & \beta_2 \\
\beta_3 & \beta_8 & \beta_7 & \beta_1 & \beta_{10} & 1 & \beta_5 & \beta_9 & \beta_4 & \beta_6 & \beta_2 & \beta_0
\end{bmatrix}
$$

$$\beta_0 \lambda = \begin{bmatrix} \beta_4 & \beta_2 & \beta_5 & \beta_{10} & \beta_9 & \beta_3 & \beta_0 & 1 & \beta_7 & \beta_{11} & \beta_6 & \beta_8 \\ \beta_2 & \beta_9 & \beta_{11} & \beta_7 & \beta_5 & \beta_8 & \beta_1 & \beta_0 & \beta_6 & \beta_3 & 1 & \beta_{10} \\ \beta_5 & \beta_{11} & \beta_8 & 1 & \beta_3 & \beta_7 & \beta_2 & \beta_4 & \beta_0 & \beta_{10} & \beta_1 & \beta_6 \\ \beta_{10} & \beta_7 & 1 & \beta_2 & \beta_6 & \beta_1 & \beta_3 & \beta_{11} & \beta_9 & \beta_0 & \beta_5 & \beta_4 \\ \beta_9 & \beta_5 & \beta_3 & \beta_6 & \beta_{11} & \beta_{10} & \beta_4 & \beta_1 & 1 & \beta_8 & \beta_0 & \beta_7 \\ \beta_3 & \beta_8 & \beta_7 & \beta_1 & \beta_{10} & 1 & \beta_5 & \beta_9 & \beta_4 & \beta_6 & \beta_2 & \beta_0 \\ \beta_0 & \beta_1 & \beta_2 & \beta_3 & \beta_4 & \beta_5 & \beta_6 & \beta_7 & \beta_8 & \beta_9 & \beta_{10} & \beta_{11} \\ 1 & \beta_0 & \beta_4 & \beta_{11} & \beta_1 & \beta_9 & \beta_7 & \beta_{10} & \beta_3 & \beta_2 & \beta_8 & \beta_5 \\ \beta_7 & \beta_6 & \beta_0 & \beta_9 & 1 & \beta_4 & \beta_8 & \beta_3 & \beta_5 & \beta_1 & \beta_{11} & \beta_2 \\ \beta_{11} & \beta_3 & \beta_{10} & \beta_0 & \beta_8 & \beta_6 & \beta_9 & \beta_2 & \beta_1 & \beta_7 & \beta_4 & 1 \\ \beta_6 & 1 & \beta_1 & \beta_5 & \beta_0 & \beta_2 & \beta_{10} & \beta_8 & \beta_{11} & \beta_4 & \beta_3 & \beta_9 \\ \beta_8 & \beta_{10} & \beta_6 & \beta_4 & \beta_7 & \beta_0 & \beta_{11} & \beta_5 & \beta_2 & 1 & \beta_9 & \beta_1 \end{bmatrix}$$

References

1. Agnew, G.B., Beth, T., Mullin, R.C., Vanstone, S.A.: Arithmetic operations in $GF(2^m)$. J. Cryptol. **6**(1), 3–13 (1993)
2. Agnew, G.B., Mullin, R.C., Onyszchuk, I., Vanstone, S.A.: An implementation for a fast public-key cryptosystem. J. Cryptol. **3**(2), 63–79 (1991)
3. Agnew, G.B., Mullin, R.C., Vanstone, S.A.: An implementation of elliptic curve cryptosystems over $F_{2^{155}}$. IEEE J. Sel. Areas Commun. **11**(5), 804–813 (1993)
4. Blake, I., Seroussi, G., Smart, N.: Elliptic Curves in Cryptography. Cambridge University Press, Cambridge (1999)
5. Erdem, S.S., Yanık, T., Koç, Ç.K.: Polynomial basis multiplication in GF(2^m). Acta Applicandae Mathematicae **93**(1–3), 33–55 (2006)
6. Gao, S.: Normal bases over finite fields. Ph.D. thesis, University of Waterloo (1993)
7. Gao, S., Lenstra Jr., H.W.: Optimal normal bases. Des. Codes Cryptgr. **2**(4), 315–323 (1992)
8. von zur Gathen, J., Shokrollahi, M.A., Shokrollahi, J.: Efficient multiplication using type 2 optimal normal bases. In: Carlet, C., Sunar, B. (eds.) WAIFI 2007. LNCS, vol. 4547, pp. 55–68. Springer, Heidelberg (2007)
9. Halbutoğulları, A., Koç, Ç.K.: Mastrovito multiplier for general irreducible polynomials. IEEE Trans. Comput. **49**(5), 503–518 (2000)
10. Hasan, M.A., Wang, M.Z., Bhargava, V.K.: Modular construction of low complexity parallel multipliers for a class of finite fields $GF(2^m)$. IEEE Trans. Comput. **41**(8), 962–971 (1992)
11. Itoh, T., Tsujii, S.: A fast algorithm for computing multiplicative inverses in $GF(2^m)$ using normal bases. Inf. Comput. **78**(3), 171–177 (1988)
12. Itoh, T., Tsujii, S.: Structure of parallel multipliers for a class of finite fields $GF(2^m)$. Inf. Comput. **83**, 21–40 (1989)
13. Koç, Ç.K., Acar, T.: Montgomery multiplication in GF(2^k). Des. Codes Cryptgr. **14**(1), 57–69 (1998)
14. Koç, Ç.K., Acar, T., Kaliski Jr., B.S.: Analyzing and comparing Montgomery multiplication algorithms. IEEE Micro **16**(3), 26–33 (1996)

15. Mastrovito, E.D.: VLSI architectures for multiplication over finite field GF(2^m). In: Mora, T. (ed.) AAECC-6. LNCS, vol. 357, pp. 297–309. Springer, Heidelberg (1988)

16. Mastrovito, E.D.: VLSI architectures for computation in Galois fields. Ph.D. thesis, Linköping University, Department of Electrical Engineering, Linköping, Sweden (1991)

17. Montgomery, P.L.: Modular multiplication without trial division. Math. Comput. **44**(170), 519–521 (1985)

18. Mullin, R., Onyszchuk, I., Vanstone, S., Wilson, R.: Optimal normal bases in $GF(p^n)$. Discrete Appl. Math. **22**, 149–161 (1988)

19. Omura, J., Massey, J.: Computational method and apparatus for finite field arithmetic (May 1986). U.S. Patent Number 4,587,627

20. Paar, C.: A new architecture for a parallel finite field multiplier with low complexity based on composite fields. IEEE Trans. Comput. **45**(7), 856–861 (1996)

21. Reyhani-Masoleh, A., Hasan, M.A.: A new construction of Massey-Omura parallel multiplier over GF(2^m). IEEE Trans. Comput. **51**(5), 511–520 (2001)

22. Saldamlı, G.: Spectral modular arithmetic. Ph.D. thesis, Oregon State University (2005)

23. Saldamlı, G., Baek, Y.J., Koç, Ç.K.: Spectral modular arithmetic for binary extension fields. In: The 2011 International Conference on Information and Computer Networks (ICICN), pp. 323–328 (2011)

24. Seroussi, G.: Table of low-weight binary irreducible polynomials (August 1998). Hewlett-Packard, HPL-98-135

25. Silverman, J.H.: Fast multiplication in finite fields GF(2^n). In: Koç, Ç.K., Paar, C. (eds.) CHES 1999. LNCS, vol. 1717, pp. 122–134. Springer, Heidelberg (1999)

26. Sunar, B., Koç, Ç.K.: Mastrovito multiplier for all trinomials. IEEE Trans. Comput. **48**(5), 522–527 (1999)

27. Wu, H., Hasan, M.A.: Low complexity bit-parallel multipliers for a class of finite fields. IEEE Trans. Comput. **47**(8), 883–887 (1998)

28. Zhang, T., Parhi, K.K.: Systematic design of original and modified Mastrovito multipliers for general irreducible polynomials. IEEE Trans. Comput. **50**(7), 734–749 (2001)

Second Invited Talk

Open Questions on Nonlinearity
and on APN Functions

Claude Carlet[1,2]([⊠])

[1] LAGA, Universities of Paris 8 and Paris 13; CNRS, UMR 7539,
Saint-Denis Cedex 02, France
[2] Department of Mathematics, University of Paris 8, 2 rue de la Liberté,
93526 Saint-Denis Cedex 02, France
claude.carlet@univ-paris8.fr

Abstract. In a first part of the paper, we recall some known open
questions on the nonlinearity of Boolean and vectorial functions and
on the APN-ness of vectorial functions. All of them have been extensively searched and seem quite difficult. We also indicate related less
well-known open questions. In the second part of the paper, we introduce four new open problems (leading to several related sub-problems)
and the results which lead to them. Addressing these problems may be
less difficult since they have not been much worked on.

Keywords: Cryptography · Boolean function · Nonlinearity · Almost
Perfect Nonlinear · Almost Bent

1 Introduction

The study of Boolean functions for cryptography has become nowadays an
important domain of research. Its connections with coding theory, sequences,
designs, difference sets, group rings, combinatorics and other domains are multiple. The width of the area is regularly increasing, around the well established
topics of Boolean functions for stream ciphers, of bent functions, and of vectorial
functions for block ciphers (S-boxes). The study of Boolean functions for stream
ciphers includes that of several parameters like algebraic degree, nonlinearity,
resiliency, and the more recent algebraic immunity and fast algebraic immunity,
which give selection criteria for the combiner and filter models in pseudo-random
generators. Since these criteria cannot be satisfied in the same time at the highest possible levels, the best possible tradeoffs between these criteria need to be
evaluated. Each time an important criterion appears (like algebraic immunity
ten years ago), the situation for searching good tradeoffs changes significantly
and can even need to be completely reconsidered (this has been the case when
came the algebraic attacks). A list of open problems on all criteria and their
tradeoffs would need more than one paper. In the present paper, we mainly
focus on open questions related to the nonlinearity.

© Springer International Publishing Switzerland 2015
Ç. Koç et al. (Eds.): WAIFI 2014, LNCS 9061, pp. 83–107, 2015.
DOI: 10.1007/978-3-319-16277-5_5

The same situation occurs for vectorial functions as for Boolean functions. However, no mandatory "logical" criterion[1] appeared since the early 90's. The three main parameters quantifying criteria are the algebraic degree, the differential uniformity and the nonlinearity. Nevertheless, side channel attacks have changed the situation of the design of best possible S-boxes, since the implementation of S-boxes by devices leaking information on the data handled inside the algorithms needs to include counter-measures to these attacks; hence the situation has also evolved for this topic. In this paper, we focus on open problems on nonlinearity and APN-ness (related to the "black box" linear and differential attacks).

Boolean functions play also a role in authentication schemes and secret sharing.

The number of researchers interested in the domain is continuously increasing. With no surprise, the open problems arising from this research are at least as deep and numerous as the results obtained. A whole book could be devoted to them and we then have here to make choices.

The purpose of the present paper is to give an idea of the main open problems on the nonlinearity of Boolean and vectorial functions and on the APN-ness of vectorial functions (we tried to choose the main important ones, but of course our choice may seem subjective) and to introduce a few new ones, which are related to new results, that we introduce as well.

2   Preliminaries

We recall in this section the definitions which will be useful at several places of the paper. Other more specific definitions will be introduced in the subsections addressing them. A *Boolean function* $f : \mathbb{F}_2^n \mapsto \mathbb{F}_2$ (in other terms, *an n-variable Boolean function*) is considered as nonlinear if it has a behavior far from that of *affine functions* (i.e. of sums of a linear form over the $\mathbb{F}_2$-vectorspace $\mathbb{F}_2^n$ and of a constant of $\mathbb{F}_2$), in some quantifiable way having cryptographic significance. A possible way for behaving differently from affine functions is to have large *algebraic degree*, since the complexity of at least three attacks (the Berlekamp-Massey attack [31], the fast algebraic attack[2] [17] and the Rønjom-Helleseth attack [34]) depends on this parameter. The algebraic degree of a Boolean function is the global degree of its unique multivariate polynomial representation called its *Algebraic Normal Form* (ANF):

$$f(x_1, \cdots, x_n) = \sum_{I \subseteq \{1,...,n\}} a_I \left(\prod_{i \in I} x_i \right) ; \; x_i, a_I \in \mathbb{F}_2 \text{ (sum to be taken modulo 2)}.$$

---

[1] By that, we mean a criterion related to those attacks on the cryptosystem viewed as a black box, by opposition to the attacks exploiting leaks, like side channel attacks.

[2] The precise parameter allowing to quantify the complexity of this attack is the fast algebraic immunity, introduced in a preliminary version of the paper [30] and studied in [16], but a large algebraic degree is a necessary condition for the resistance to the fast algebraic attack.

But the name of *nonlinearity* has been historically given to a parameter, denoted ny $nl(f)$, which happened later to be related to the attack by linear approximation [27] and to the fast correlation attack [18]. This parameter is equal to the minimum *Hamming distance* $d(f, \ell) = |\{x \in \mathbb{F}_2^n \,;\, f(x) \neq \ell(x)\}|$ (where $|\ldots|$ denotes the cardinality) from the function to all affine functions ℓ. Note that, contrary to what suggests this term, having a high nonlinearity is not sufficient for f to be highly nonlinear: f needs also to have several other properties, including a large algebraic degree, a large algebraic immunity, a large fast algebraic immunity... See more in [7].

The nonlinearity can be studied through the *Walsh transform* of f, defined as:

$$W_f(a) = \sum_{x \in \mathbb{F}_2^n} (-1)^{f(x) + a \cdot x} \text{ (sum in } \mathbb{Z}) \tag{1}$$

where $a \cdot x$ is some inner product in $\mathbb{F}_2^n$ (changing the inner product only permutes the values of the Walsh transform), for instance the usual inner product in $\mathbb{F}_2^n$: $a \cdot x = \sum_{i=1}^{n} a_i x_i$ (sum to be taken modulo 2) or, if $\mathbb{F}_2^n$ is endowed with the structure of the field $\mathbb{F}_{2^n}$, the usual inner product in $\mathbb{F}_{2^n}$: $a \cdot x = tr_n(ax)$ where $tr_n(x) = \sum_{i=0}^{n-1} x^{2^i}$ is the trace function from $\mathbb{F}_{2^n}$ to $\mathbb{F}_2$. The relation is:

$$nl(f) = 2^{n-1} - \frac{1}{2} \max_{a \in \mathbb{F}_2^n} |W_f(a)|, \tag{2}$$

where $|\ldots|$ denotes the absolute value. This relation together with the *Parseval relation*:

$$\sum_{a \in \mathbb{F}_2^n} W_f^2(a) = 2^{2n} \tag{3}$$

(which is valid for every Boolean function f) implies the so-called covering radius bound $nl(f) \leq 2^{n-1} - 2^{n/2-1}$. A function f is called *bent* [35] if it achieves this bound with equality. Note that n must then be even. It is proved in [35] that any bent function has algebraic degree at most $n/2$. If f is an n-variable bent function and A is an affine permutation of $\mathbb{F}_2^n$, then $f \circ A$ is bent. The equivalence relation between such f and $f \circ A$ is called *affine equivalence*. This relation is extended into another equivalence relation called *Extended Affine equivalence* (EA-equivalence) by possibly adding an affine Boolean function to $f \circ A$.

The Walsh transform is closely related to the so-called Fourier transform $\widehat{f}(a) := \sum_{x \in \mathbb{F}_2^n} f(x)(-1)^{x \cdot a}$ (sum in $\mathbb{Z}$), since for every Boolean function f, we have:

$$W_f(a) = -2\widehat{f}(a) \text{ for } a \neq 0 \tag{4}$$

and

$$W_f(0) = 2^n - 2\widehat{f}(0) = 2^n - 2w_H(f), \tag{5}$$

where w_H denotes the Hamming weight.

The nonlinearity $nl(F)$ of a vectorial function $F : \mathbb{F}_2^n \mapsto \mathbb{F}_2^m$ (called an (n, m)-function) equals by definition the minimum nonlinearity of the linear combinations of its coordinate functions with non-all-zero coefficients, that is, of the Boolean functions $v \cdot F$, $v \in \mathbb{F}_2^m$, $v \neq 0$ (these functions are called the *component functions* of F; their set is independent of the choice of the inner product "."). Of course we have here again the covering radius bound $nl(F) \leq 2^{n-1} - 2^{n/2-1}$ and there is equality if and only if each component function of F is bent (we say then that F is bent or *perfect nonlinear*). Bent (n, m)-functions exist only for n even and $m \leq n/2$ as proved by Nyberg [32]. Another bound called the *Sidelnikov-Chabaud-Vaudenay* (SCV) bound exists for $m \geq n-1$ (its general expression involves a square root and the expression under this square root is non-negative under this condition). For $m = n-1$, it coincides with the covering radius bound, and for $m \geq n$ it improves upon it. For $m = n$, it states that $nl(F) \leq 2^{n-1} - 2^{\frac{n-1}{2}}$. An (n, n)-function achieving this bound with equality is called *Almost Bent* (in brief, AB; necessarily, n is odd).

Any AB function is Almost Perfect Nonlinear (in brief, APN), that is, is such that every equation $F(x) + F(x + a) = b$, where $a, b \in \mathbb{F}_2^n$ and $a \neq 0$, has at most 2 solutions $x \in \mathbb{F}_2^n$ (that is, has 0 or 2 solutions; equivalently). In other words, any non-trivial first-order derivative is 2-to-1. There exist APN functions in odd dimension (that is, APN (n, n)-functions with n odd) which are not AB. But if the Walsh transform of an APN function in odd dimension has values divisible by $2^{\frac{n+1}{2}}$, then it is AB.

The notion of APN function has been weakened into the notion of differentially δ-uniform (n, m)-function (while the term of APN is used only for (n, n)-functions), which corresponds to the fact that any equation $F(x) + F(x + a) = b$, $a \neq 0$ has at most δ solutions.

EA-equivalence is extended to vectorial functions as follows: two functions F and G are EA-equivalent if there exist two affine permutations A, A' and an affine function A'' such that $G = A' \circ F \circ A + A''$. A notion of equivalence which is more general than EA-equivalence exists and is now called CCZ-equivalence [12] (it could also be called graph equivalence). Two (n, m)-functions F and G are CCZ-equivalent if there exists an affine permutation of $\mathbb{F}_2^n \times \mathbb{F}_2^m$ which maps the graph $\{(x, F(x)), x \in \mathbb{F}_2^n\}$ of F to the graph of G. This more general CCZ-equivalence preserves nonlinearity and APN-ness but does not preserve the algebraic degree, contrary to EA-equivalence (when the algebraic degree is at least 2). For bent vectorial functions and for Boolean functions, CCZ-equivalence reduces to EA-equivalence (see e.g. [8]).

3   Open Problems

In this section, we give a list of well-known open questions and we describe, for some of them, related side-questions.

3.1   Investigate the Structure of Class $\mathcal{PS}$ of Bent Functions, Evaluate Its Size, Find a Constructive Definition of These Functions

There exist numerous open problems on bent functions. A quite non-exhaustive list, already pretty long, is given in [10]. An interesting open question is to:

– Know more on the structure of $\mathcal{PS}$ bent functions
– and determine whether they cover a large part of bent functions.

Defined by Dillon in [19,20], class $\mathcal{PS}$ is the union of classes $\mathcal{PS}^-$ and $\mathcal{PS}^+$, which contain respectively those n-variable Boolean functions f whose supports $supp(f) = \{x \in \mathbb{F}_2^n \,;\, f(x) = 1\}$ are the union of $2^{n/2-1}$ (resp. $2^{n/2-1}+1$) pairwise supplementary $n/2$-dimensional $\mathbb{F}_2$-vector subspaces of $\mathbb{F}_2^n$ (having trivial pairwise intersection and pairwise sum equal to $\mathbb{F}_2^n$), 0 excluded (resp. 0 included). All of these functions are bent. The functions of class $\mathcal{PS}^-$ all have algebraic degree $n/2$ while those of class $\mathcal{PS}^+$ can have lower algebraic degrees. Until recently, two sets (at least) of $2^{n/2-1}$ supplementary spaces were known, up to affine equivalence:

1. The $2^{n/2}+1$ multiplicative cosets of $\mathbb{F}_{2^{n/2}}^*$ in $\mathbb{F}_{2^n}^*$ (to each of which is of course adjoined 0); these $2^{n/2}+1$ pairwise supplementary vectorspaces completely cover $\mathbb{F}_{2^n}$; their set is then a *full spread*; the corresponding $\mathcal{PS}^-$ and $\mathcal{PS}^+$ functions obtained by choosing $2^{n/2-1}$ (resp. $2^{n/2-1}+1$) of these spaces constitute the so-called $\mathcal{PS}_{ap}$ class ("ap" for "affine plane"); all of them have algebraic degree $n/2$;
2. For $n/2$ even, a set of $2^{n/2-1}+1$ pairwise supplementary $n/2$-dimensional $\mathbb{F}_2$-vector subspaces introduced by Dillon [20] (and reported in [7]) whose corresponding $\mathcal{PS}^+$ function is quadratic (hence, up to EA-equivalence, every quadratic function belongs to $\mathcal{PS}^+$ for $n \equiv 0 \,[\mathrm{mod}\ 4]$).

Recently, Wu [38] introduced three other full spreads, deduced from the Dempwolff-Müller pre-quasifield (an Abelian finite group having a second law $*$ which is left-distributive with respect to the first law and is such that the right and left multiplications by a nonzero element are bijective, and that the left-multiplication by 0 is absorbent), the Knuth pre-semifield (a pre-quasifield which remains one when $a*b$ is replaced by $b*a$) and the Kantor pre-semifield. He determined explicitly the corresponding $\mathcal{PS}$ functions (which exist then when $n/2$ is odd, only). This makes now four potentially inequivalent classes of $\mathcal{PS}$ bent functions if $n/2$ is odd and two if $n/2$ is even. The question is to know whether $\mathcal{PS}$ functions are numerous (and we would know only a tiny part of them). This question is completely open. A way of addressing it with a negative answer would be to find a particular structure of the $\mathcal{PS}$ functions. Such question is also open.

 $\mathcal{PS}$ class has been generalized in [5] into the so-called $\mathcal{GPS}$ class, of those $\{0,1\}$-valued functions of the form $f(x) = \sum_{i=1}^k m_i 1_{E_i}(x) - 2^{n/2-1}\delta_0(x)$ where k may be any positive integer, the E_i are pairwise supplementary $n/2$-dimensional $\mathbb{F}_2$-vector subspaces of $\mathbb{F}_2^n$, 1_{E_i} is the indicator function of E_i, the m_i's are integers, and δ_0 is the Dirac symbol. The only fact that $\sum_{i=1}^k m_i 1_{E_i}(x) - 2^{n/2-1}\delta_0(x)$

takes its values in $\{0,1\}$ suffices for defining a bent function. The functions of $\mathcal{PS}$ correspond to the case where all values m_i equal 1. Guillot [26] has proved that, up to the translations $f(x) \mapsto f(x+a)$, $\mathcal{GPS}$ class covers all bent functions. Is it possible that $\mathcal{GPS}$ be so large and $\mathcal{PS}$ be however small is an open question.

Other open questions are related:

– If $\mathcal{PS}$ happens to be large, is it also the case of the subclass of those $\mathcal{PS}$ functions which are related to full spreads?

Remark

1. As observed by Dillon, the $\mathcal{PS}$ functions related to those partial spreads which can be extended to spreads of larger sizes (in particular, those related to full spreads) have necessarily algebraic degree $n/2$.
2. The number of bent functions of algebraic degree $n/2$ seems large among all bent functions.
3. What are the possible algebraic degrees of $\mathcal{PS}^+$ bent functions?

3.2   Determine Efficient Lower and Upper Bounds on the Number of n-variable Bent Functions; Clarify the Tokareva Open Question

The best known general upper bound on the number of bent functions is given in [15]. This bound seems weak when compared with the actual numbers, known for $n \leq 8$ (see [29]) and there seems to be plenty of room for improvement. However, this bound already 12 year old has not been improved yet. A first open question is then to:

– Find an efficient upper bound on the number of bent functions.

The situation with lower bounds is still worse: all that is done so far is to bound below the number of known bent functions (this number cannot be exactly evaluated because it is difficult for a given function to count the number of EA-equivalent functions, and it is still more difficult to classify the known bent functions up to EA-equivalence). There exists an open question posed by Tokareva in [37] which is related:

– For any even $n \geq 4$, every n-variable Boolean function of algebraic at most $n/2$ is it equal to the sum (mod 2) of two bent functions?

The most tempting reply to this question is "no", because bent functions seem quite rare, but Tokareva shows that the reply is "yes" for $n = 4, 6$ and for several classes of Boolean functions (see [37]). As far as we know, no specific property of the sums of two bent functions could be found yet to allow replying negatively to the question. If this question could be answered positively, the square of the number of n-variable bent functions would be larger than or equal to the number $2^{\sum_{i=0}^{n/2} \binom{n}{i}}$ of Boolean functions of algebraic degrees at most $n/2$, and this would considerably increase the best known lower bound. In any case, the question:

– Find an efficient lower bound on the number of bent functions remains quite open.

3.3 Find a Better Upper Bound Than the Covering Radius Bound for the Nonlinearity of Functions $F : \mathbb{F}_2^n \mapsto \mathbb{F}_2^m$ when n is Odd and $m < n$ or when n is Even and $n/2 < m < n$

We have seen in Sect. 2 that the covering radius bound on the nonlinearity of (n, m)-functions is tight only for n even and $m \leq n/2$. There should then be room for improvement. However, no significantly better bound than $nl(F) < 2^{n-1} - 2^{n/2-1}$ is known for general (n, m)-functions such that $m > n/2$. It turns out that the known ideas, used in the proofs of the covering radius and SCV bounds, which mainly consist in considering power moments of orders 2 and 4 of the Walsh transform, are not adaptable for improving upon the covering radius bound. In particular, considering power moments of larger orders failed until now to give an improvement. New ideas seem necessary.

3.4 Find Classes of AB/APN Functions by Using CCZ-equivalence with Kasami (resp. Welch, Niho, Dobbertin) Functions

In [4] were constructed APN (n, n)-functions (n even) and AB functions (n odd) from known ones by using CCZ-equivalence. To do so, given a function F, we need to find an affine permutation $\mathcal{L}$ of $\mathbb{F}_2^n \times \mathbb{F}_2^n$ such that, denoting the value of $\mathcal{L}(x, y)$ by $(L_1(x, y), L_2(x, y))$, where $L_1(x, y), L_2(x, y) \in \mathbb{F}_2^n$, the function $F_1(x) = L_1(x, F(x))$ is a permutation. This is a necessary and sufficient condition for the image of the graph of F by $\mathcal{L}$ to be the graph of a function. Several cases of such $\mathcal{L}$ were found for the function $F(x) = x^{2^i+1}$ where $(i, n) = 1$ (this latter condition is necessary and sufficient for such Gold function to be AB if n is odd and APN if n is even). The resulting functions happened to be EA-inequivalent to the Gold functions and more generally to any power function. The Gold functions are the only power functions for which such method based on CCZ-equivalence could be achieved (it could be applied to other quadratic functions, see Subsect. 3.7). To apply it to the other known classes of AB/APN power (non-quadratic) functions is an open problem.

Remark: For n odd, the Kasami $x^{4^k-2^k+1}$ function equals $F_2 \circ F_1^{-1}(x)$, where $F_1(x)$ and $F_2(x)$ are respectively the Gold functions x^{2^k+1} and $x^{2^{3k}+1}$. Hence, the first step would be to investigate permutations of the form $L_1^1(x^{2^k} + 1) + L_1^2(x^{2^{3k}} + 1)$, that is, to find L_1^1 and L_1^2 linear such that for every $u \neq 0$ and every x, we have $L_1^1(x^{2^k}u + xu^{2^k} + u^{2^k+1}) + L_1^2(x^{2^{3k}}u + xu^{2^{3k}} + u^{2^{3k}+1}) \neq 0$.

3.5 Find Infinite Classes of AB/APN Functions CCZ-inequivalent to Power Functions and to Quadratic Functions

We have only one APN function known (with $n = 6$) which is CCZ-inequivalent to power functions and to quadratic functions [24]. We still need to find an example of AB function having the same property. Finding infinite classes of APN (resp. AB) functions with the same property is still more open.

3.6   Determine Whether There Exist Vectorial Functions $F : \mathbb{F}_2^n \mapsto \mathbb{F}_2^n$, n Even, with Better Nonlinearity than the Multiplicative Inverse, Gold and Kasami Functions in Even Dimension

There is a big gap between the best possible nonlinearity $2^{n-1} - 2^{\frac{n-1}{2}}$ of (n,n)-functions for n odd, achieved by AB functions, and the best known nonlinearity $2^{n-1} - 2^{n/2}$ of (n,n)-functions for n even, which is achieved by the Gold APN functions already recalled, the Kasami APN functions $x^{4^i - 2^i + 1}$ where $(i,n) = 1$, and by the multiplicative inverse function $x^{2^n - 2}$. The gap could seem not so important, but it is, since what matters for the complexity of attacks by linear approximation is not the value of $nl(F)$ but the value of $\frac{2^{n-1} - nl(F)}{2^{n-1}}$. Finding functions with better nonlinearity (and still more relevantly to cryptography, with better nonlinearity and good differential uniformity) or proving that such function does not exist is an open question.

3.7   Find APN Permutations in Even Dimension at Least 8

This question is considered as the big open problem on vectorial functions for cryptography. During a long time, it has been conjectured that no APN (n,n)-permutation exists for n even. Then Browning, Dillon, McQuistan and Wolfe obtained in [2] an APN $(6,6)$-permutation by applying CCZ-equivalence to the so-called Kim function (i.e. the univariate polynomial function over $\mathbb{F}_{2^6}$ first presented in [1] and equal to $X^3 + X^{10} + \alpha X^{24}$, where α is a primitive element of $\mathbb{F}_{2^6}$). The open question is now to obtain one in at least 8 variables (exactly 8 variables would be the most interesting for applications, but it may be possible that $n \equiv 2 \pmod 4$ be necessary) or better an infinite class. For reasons that we shall recall below, a way for this could be to find generalizations of the Kim function. Faruk Gologlu [25] has proved that function $F(x) = x^{2^k + 1} + (tr_{n/2}^n(x))^{2^k + 1}$ is APN for n divisible by 4, where $(n,k) = 1$ and $tr_{n/2}^n(z) = z + z^q$, $q = 2^{n/2}$, is the trace function from $\mathbb{F}_{2^n}$ to $\mathbb{F}_{2^{n/2}}$. Unfortunately, this function does not seem to lead to a permutation. A related question is then to:

- Find more functions or, better, infinite classes of functions of the form $X^3 + aX^{2+q} + bX^{2q+1} + cX^{3q}$ where $q = 2^{n/2}$, or more generally of the form $X^{2^k+1} + aX^{2^k+q} + bX^{2^k q+1} + cX^{2^k q+q}$, where $(k,n) = 1$ (we shall call "generalized Kim" such APN functions).

This would maybe lead to an APN permutation. Indeed, the method used in [2] (and revisited in [25]) is based on the fact that an (n,n)-function F is CCZ-equivalent to a permutation if and only if there exist in $\mathbb{F}_{2^n} \times \mathbb{F}_{2^n}$ two supplementary n-dimensional vector spaces (that is, given their dimension, two spaces having trivial intersection) on which the Walsh transform of F vanishes (except at 0). Indeed, there exists then a linear permutation, mapping $\mathbb{F}_{2^n} \times \{0\}$ and $\{0\} \times \mathbb{F}_{2^n}$ to these two spaces respectively, and this directly leads to $\mathcal{L}$ such that the resulting CCZ-equivalent function is bijective. The generalized Kim functions (of the form $F(X) = X^{2^k+1} + aX^{2^k+q} + bX^{2^k q+1} + cX^{2^k q+q}$ where $(k,n) = 1$) have the peculiarity that $F(yX) = y^{2^k+1}F(X)$ for every $y \in \mathbb{F}_{2^n}$. This implies that the Walsh transform $W_F(a,u) = \sum_{x \in \mathbb{F}_{2^n}} (-1)^{tr_n(uF(x)+ax)}$ of F satisfies $W_F(a,u) = W_F(ay, uy^{2^k+1})$ for every $y \in \mathbb{F}_{2^n}^*$, which can increase the possibility of finding n-dimensional subspaces as indicated above.

The Browning et al. method as suggested (in a less general but more constructive way) at the end of paper [2] can be seen as follows in bivariate form. We write $X = x + wy$ where $x, y \in \mathbb{F}_{2^m}$ where $m = n/2$, and $w \in \mathbb{F}_{2^n} \setminus \mathbb{F}_{2^m}$. We have $X^{2^k+1} = (x+wy)^{2^k}(x+wy) = x^{2^k+1} + wx^{2^k}y + w^{2^k}xy^{2^k} + w^{2^k+1}y^{2^k+1}$, $X^{2^k+q} = (x+wy)^{2^k}(x+wy)^q = x^{2^k+1} + w^q x^{2^k}y + w^{2^k}xy^{2^k} + w^{2^k+q}y^{2^k+1}$, $X^{2^k q+1} = (x+wy)^{2^k q}(x+wy) = x^{2^k+1} + wx^{2^k}y + w^{2^k q}xy^{2^k} + w^{2^k q+1}y^{2^k+1}$ and $X^{2^k q+q} = (X^{2^k+1})^q = x^{2^k+1} + w^q x^{2^k}y + w^{2^k q}xy^{2^k} + w^{2^k q+q}y^{2^k+1}$. Hence, the generalized Kim function $X^{2^k+1} + aX^{2^k+q} + bX^{2^k q+1} + cX^{2^k q+q}$ equals

$$
\begin{aligned}
K(x,y) := & (1 + a + b + c)x^{2^k+1} + (w + aw^q + bw + cw^q)x^{2^k}y \\
& + (w^{2^k} + aw^{2^k} + bw^{2^k q} + cw^{2^k q})xy^{2^k} \\
& + (w^{2^k+1} + aw^{2^k+q} + bw^{2^k q+1} + cw^{2^k q+q})y^{2^k+1}.
\end{aligned}
$$

If we translate the construction of APN permutation of [2] in the bivariate setting, the output of the generalized Kim function being expressed in the form $(F(x,y), G(x,y))$, it is sufficient for constructing the permutation by CCZ equivalence, that the functions $(x,y) \to (x, F(x,y))$ and $(x,y) \to (y, G(x,y))$ be permutations. Indeed, we introduce then the linear permutation of $(\mathbb{F}_{2^m} \times \mathbb{F}_{2^m})^2$:

$$
\mathcal{L}((x,y),(x',y')) = (L_1((x,y),(x',y')), L_2((x,y),(x',y'))) = ((x,x'),(y,y')).
$$

Since the functions $F_1(x,y) := L_1((x,y), K(x,y)) = (x, F(x,y))$ and $F_2(x,y) := L_2((x,y), K(x,y)) = (y, G(x,y))$ are permutations of $\mathbb{F}_{2^m} \times \mathbb{F}_{2^m}$, the function $F_2 \circ F_1^{-1}$ is a permutation CCZ-equivalent to the generalized Kim function (and is therefore APN). Note that the double condition that the functions $(x,y) \to (x, F(x,y))$ and $(x,y) \to (y, G(x,y))$ are permutations is equivalent to the facts that, for every $x \in \mathbb{F}_{2^m}$, the function $y \in \mathbb{F}_{2^m} \to F(x,y)$ is a permutation of $\mathbb{F}_{2^m}$ and for every $y \in \mathbb{F}_{2^m}$, the function $x \in \mathbb{F}_{2^m} \to G(x,y)$ is a permutation as well.

Let us show now how such representation $(F(x,y), G(x,y))$ of the generalized Kim function can be found. Functions G and F are of the form

$$
\begin{aligned}
tr_m^n(\beta K(x,y)) = & \\
(\beta + a\beta + b\beta + c\beta + \beta^q & + a^q\beta^q + b^q\beta^q + c^q\beta^q)x^{2^k+1} \\
+ (w\beta + aw^q\beta + bw\beta + cw^q\beta & + w^q\beta^q + a^q w\beta^q + b^q w^q\beta^q + c^q w\beta^q)x^{2^k}y \\
+ (w^{2^k}\beta + aw^{2^k}\beta + bw^{2^k q}\beta & + cw^{2^k q}\beta + w^{2^k q}\beta^q + a^q w^{2^k q}\beta^q + b^q w^{2^k}\beta^q + \\
& c^q w^{2^k}\beta^q)xy^{2^k} \\
+ (w^{2^k+1}\beta + aw^{2^k+q}\beta + bw^{2^k q+1}\beta & + cw^{2^k q+q}\beta + w^{2^k q+q}\beta^q + a^q w^{2^k q+1}\beta^q + \\
& b^q w^{2^k+q}\beta^q + c^q w^{2^k+1}\beta^q)y^{2^k+1}.
\end{aligned}
$$

and $tr_m^n(\gamma K(x,y))$, where β, γ belong to $\mathbb{F}_{2^n}$ and are not in the same multiplicative coset of $F_{2^m}^*$ (that is, are not F_{2^m} linearly dependent). A function of the form $ax^{2^k+1} + bx^{2^k} + cx + d$ is a permutation of $\mathbb{F}_{2^m}$ (for m odd) under the sufficient condition that ($a = b = 0$ and $c \neq 0$) or ($a = c = 0$ and $b \neq 0$) or ($a \neq 0$ and $ab^{2^k} = a^{2^k}c$): indeed the two first cases are straightforward and, in the third case, the function equals then $a(x + b/a)^{2^k+1}$ plus a constant and $x \rightarrow x^{2^k+1}$ is a permutation of $\mathbb{F}_{2^m}$ for m odd. In fact, it can be proved that this condition is also necessary for every odd m, since the function $x^{2^k+1} + \lambda x$ is a permutation of $\mathbb{F}_{2^m}$ if and only if the Boolean function $tr_m(\mu x^{2^k+1} + \mu \lambda x)$ is balanced for every $\mu \neq 0$; this Boolean function has linear kernel of equation $\mu x^{2^k} + (\mu x)^{2^{n-k}} = 0$, which is equivalent to $\mu x^{2^k+1} \in \mathbb{F}_2$ and the condition becomes that $tr\left(1 + \frac{\mu\lambda}{2^k + \sqrt[1]{\mu}}\right) = 0$ for every $\mu \neq 0$ and this is equivalent to $\lambda = 0$ since $tr_m(1) = 1$ and $\frac{\mu}{2^k + \sqrt[1]{\mu}} = \left(\sqrt[2^k+1]{\mu}\right)^{2^k}$ ranges over $\mathbb{F}_{2^m}$ when μ does. We want to characterize the conditions under which the function $x \rightarrow tr_m^n(\beta K(x,y))$ is a permutation for every y. Clearly for $y = 0$, the condition is $\beta + a\beta + b\beta + c\beta + \beta^q + a^q\beta^q + b^q\beta^q + c^q\beta^q \neq 0$, that is, $\beta + a\beta + b\beta + c\beta \notin \mathbb{F}_{2^m}$ and, for $y \neq 0$, the function is then bijective if and only if $\beta + a\beta + b\beta + c\beta \notin \mathbb{F}_{2^m}$ and $(\beta + a\beta + b\beta + c\beta + \beta^q + a^q\beta^q + b^q\beta^q + c^q\beta^q)(w\beta + aw^q\beta + bw\beta + cw^q\beta + w^q\beta^q + a^qw\beta^q + b^qw^q\beta^q + c^qw\beta^q)^{2^k} = (\beta + a\beta + b\beta + c\beta + \beta^q + a^q\beta^q + b^q\beta^q + c^q\beta^q)^{2^k}(w^{2^k}\beta + aw^{2^k}\beta + bw^{2^kq}\beta + cw^{2^kq}\beta + w^{2^kq}\beta^q + a^qw^{2^kq}\beta^q + b^qw^{2^k}\beta^q + c^qw^{2^k}\beta^q)$. Note that y does not appear in these equalities because it comes as a product of each part with the same power and can then be eliminated. This is how the peculiarity of the generalized Kim function observed by Browning et al. explains why the Kim function worked so well. Similarly, the function $y \rightarrow tr_m^n(\gamma K(x,y))$ is a permutation for every x if and only if $w^{2^k+1}\gamma + aw^{2^k+q}\gamma + bw^{2^kq+1}\gamma + cw^{2^kq+q}\gamma \notin \mathbb{F}_{2^m}$ and $(w^{2^k+1}\gamma + aw^{2^k+q}\gamma + bw^{2^kq+1}\gamma + cw^{2^kq+q}\gamma + w^{2^kq+q}\gamma^q + a^qw^{2^kq+1}\gamma^q + b^qw^{2^k+q}\gamma^q + c^qw^{2^k+1}\gamma^q)(w^{2^k}\gamma + aw^{2^k}\gamma + bw^{2^kq}\gamma + cw^{2^kq}\gamma + w^{2^kq}\gamma^q + a^qw^{2^kq}\gamma^q + b^qw^{2^k}\gamma^q + c^qw^{2^k}\gamma^q)^{2^k} = (w^{2^k+1}\gamma + aw^{2^k+q}\gamma + bw^{2^kq+1}\gamma + cw^{2^kq+q}\gamma + w^{2^kq+q}\gamma^q + a^qw^{2^kq+1}\gamma^q + b^qw^{2^k+q}\gamma^q + c^qw^{2^k+1}\gamma^q)^{2^k}(w\gamma + aw^q\gamma + bw\gamma + cw^q\gamma + w^q\gamma^q + a^qw\gamma^q + b^qw^q\gamma^q + c^qw\gamma^q)$. A computer search allows then hopefully to find the desired APN permutation.

3.8   Find Secondary Constructions of APN or Differentially 4-uniform S-boxes

It is also a big problem on APN functions to find constructions of such functions from functions in smaller dimensions. If it could be solved, this would considerably ease the research of APN functions (and of S-boxes for block ciphers). The existence of secondary constructions of Boolean functions (that is, single-output functions) has allowed to construct many highly nonlinear (e.g. bent) functions in the last decades. The difficulty in the case of APN functions, because these functions are in the same number of input and output bits, is to increase in the same time the length of the input and the length of the output, while ensuring the APN property. Another approach has been used in [9]. It consists in building (n, n)-functions by concatenating two $(n, n/2)$-functions F and G (for n even). The interest of such parameters is that they give the possibility of choosing one of the functions (or both) bent. The system of equations expressing that a non-trivial derivative of the APN function is 2-to-1 is

$$\begin{cases} F(x, y + F(x + a, y + b) = c \\ G(x, y) + G(x + a, y + b) = d \end{cases}. \text{ The first equation has exactly } 2^{n/2} \text{ solutions,}$$

for every $(a, b) \neq (0, 0)$ and every c. This potentially eases the research of G such that (F, G) is APN. This has led to new APN and differentially 4-uniform functions in [9]. One of these differentially 4-uniform functions has been used for the S-box of a new block cipher called PICARO [33]. Its very simple algebraic expression allowed to implement counter-measures to side channel attacks with minimized cost. Note that, since a bent function is never balanced, the (n, n)-function resulting from such concatenation of a bent function F with another function G cannot be bijective. A secondary construction allowing to construct APN or differentially 4-uniform permutations would be of course much nicer.

3.9 Determine Whether APN Functions Can Have Low Nonlinearity

All the APN functions whose nonlinearities could be studied so far seem to have rather good nonlinearity. For n odd, as recalled by A. Canteaut in her "Habilitation à Diriger des Recherches", the APN functions whose extended Walsh spectra (i.e. distributions of the Walsh transform absolute values) could be studied have three possible extended Walsh spectra: the spectrum of the AB functions (e.g. the Gold functions) which gives a nonlinearity of $2^{n-1} - 2^{\frac{n-1}{2}}$, the spectrum of the inverse function, which takes any value divisible by 4 in the range $]-2^{n/2+1}; 2^{n/2+1}[$ and gives a nonlinearity close to $2^{n-1} - 2^{n/2}$ and the spectrum of the Dobbertin function $F(x) = x^{2^{\frac{4n}{5}} + 2^{\frac{3n}{5}} + 2^{\frac{2n}{5}} + 2^{\frac{n}{5}} - 1}$, $x \in \mathbb{F}_{2^n}$, with n divisible by 5, which is more complex (it is divisible by $2^{n/5}$ and not divisible by $2^{2n/5+1}$) but is unknown for large n. For n even, the spectra are those of the Gold functions and of the Dobbertin function and of one exceptional quadratic APN function for $n = 6$ found in [1]. But the nonlinearity of the known (infinite class of) APN functions which is the worst, namely the Dobbertin function, is not known. So the good nonlinearity of known APN functions is only conjectured. More problematically, the only lower bound valid for all APN functions (known or not) which could be proved so far is $nl(F) > 0$, see [8]. Is it that the found APN functions are not representative of the whole class or that we just were not able to prove a better bound?

It is possible to say more about those APN functions whose extended Walsh spectra are sparse around $2^{n/2}$. But this is of course a partial result.

Proposition 1. *Let F be an APN function in $n > 2$ variables. For every real numbers a and b such that $a \leq b$, let $N_{a,b}$ be the number of ordered pairs $(u, v) \in$*

$F_2^n \times (F_2^n \setminus \{0\})$ *such that* $W_{v \cdot F}^2(u) \in]2^n + a; 2^n + b[$. *Then the nonlinearity of F is lower bounded by*

$$2^{n-1} - \frac{1}{2}\sqrt{2^n + \frac{1}{2}(b + a + \sqrt{\Delta_{a,b}})},$$

where $\Delta_{a,b} = (N_{a,b} + 1)(b - a)^2 + a\,b\,2^{n+2}(2^n - 1) + 2^{4n+2} - 2^{3n+2}$.

Proof: We have $\sum_{u \in F_2^n, v \in F_2^n \setminus \{0\}} W_{v \cdot F}^4(u) = 2^{3n+1}(2^n - 1)$ (this condition is necessary and sufficient for F being APN, see e.g. [8]), and it is easily deduced that $\sum_{\substack{u \in F_2^n, \\ v \in F_2^n \setminus \{0\}}} (W_{v \cdot F}^2(u) - 2^n)^2 = 2^{4n} - 2^{3n}$. We have then $\sum_{\substack{u \in F_2^n, \\ v \in F_2^n \setminus \{0\}}} (W_{v \cdot F}^2(u) - 2^n - a)(W_{v \cdot F}^2(u) - 2^n - b) = 2^{4n} - 2^{3n} + a\,b\,2^n(2^n - 1)$, since $\sum_{u \in F_2^n, v \in F_2^n \setminus \{0\}} (W_{v \cdot F}^2(u) - 2^n) = 0$. Since the expression $(x - a)(x - b)$ takes its minimum at $x = \frac{b+a}{2}$ and this minimum is $-\frac{(b-a)^2}{4}$, we have $(W_{v \cdot F}^2(u) - 2^n - a)(W_{v \cdot F}^2(u) - 2^n - b) \geq -\frac{(b-a)^2}{4}$ for the $N_{a,b}$ ordered pairs such that $W_{v \cdot F}^2(u) \in]2^n + a; 2^n + b[$ and $(W_{v \cdot F}^2(u) - 2^n - a)(W_{v \cdot F}^2(u) - 2^n - b) \geq 0$ for all the others. Considering the sum of the non-negative numbers equal to $(W_{v \cdot F}^2(u) - 2^n - a)(W_{v \cdot F}^2(u) - 2^n - b) + \frac{(b-a)^2}{4}$ if $W_{v \cdot F}^2(u) \in]2^n + a; 2^n + b[$ and to $(W_{v \cdot F}^2(u) - 2^n - a)(W_{v \cdot F}^2(u) - 2^n - b)$ otherwise, and using that an element of a positive sequence is smaller than or equal to the sum of the elements of the sequence, we deduce that $(W_{v \cdot F}^2(u) - 2^n - a)(W_{v \cdot F}^2(u) - 2^n - b) \leq 2^{4n} - 2^{3n} + a\,b\,2^n(2^n - 1) + N_{a,b}\frac{(b-a)^2}{4}$ for any $(u, v) \in F_2^n \times (F_2^n \setminus \{0\})$, that is, $(W_{v \cdot F}^2(u) - 2^n)^2 - (b+a)(W_{v \cdot F}^2(u) - 2^n) + ab - (2^{4n} - 2^{3n} + a\,b\,2^n(2^n - 1) + N_{a,b}\frac{(b-a)^2}{4}) \leq 0$, which implies

$$\frac{1}{2}(b + a - \sqrt{\Delta_{a,b}}) \leq W_{v \cdot F}^2(u) - 2^n \leq \frac{1}{2}(b + a + \sqrt{\Delta_{a,b}}),$$

where $\Delta_{a,b} = (b + a)^2 - 4(ab - 2^{4n} + 2^{3n} - a\,b\,2^n(2^n - 1) - N_{a,b}\frac{(b-a)^2}{4}) = (N_{a,b} + 1)(b - a)^2 + a\,b\,2^{n+2}(2^n - 1) + 2^{4n+2} - 2^{3n+2}$. Relation (2) applied to $f = v \cdot F$ completes the proof. $\square$

Proposition 1 implies the known result that if $W_{v \cdot F}^2(u)$ does not take values in the range $]0; 2^{n+1}[$, then (taking $b = -a = 2^n$ and $N_{a,b} = 0$), the nonlinearity is lower bounded by $2^{n-1} - 2^{\frac{n-1}{2}}$, that is, F is AB (which shows that the bound of Proposition 1 is tight). More generally, if we take $ab = -2^{2n}$ and $N_{a,b} = 0$, we have $\Delta_{a,b} = (b - a)^2$ and we obtain:

Corollary 1. *Let F be an APN function and let b be a positive number. If $W_{v \cdot F}^2(u)$ does not take values in the range $]2^n - \frac{2^{2n}}{b}; 2^n + b[$ when $u \in F_2^n$ and $v \in F_2^n \setminus \{0\}$, the nonlinearity of F is lower bounded by $2^{n-1} - \frac{1}{2}\sqrt{2^n + b}$.*

Clearly, because of the Sidelnikov-Chabaud-Vaudenay bound, the conclusion, and therefore the assumption, is possible only if $b \geq 2^n$.

3.10 Determine Whether Differentially 6-uniform $(n, n - 2)$-functions Exist

According to Nyberg's results [32], for $n \geq 5$, the value of the differential uniformity $\delta = \max \left(|\{x \in \mathbb{F}_2^n \,;\, F(x) + F(x + a) = b\}| \,;\, a \in \mathbb{F}_2^n, b \in \mathbb{F}_2^{n-2}, a \neq 0 \right)$ of an $(n, n - 2)$-function F is bounded below by 6, since the differential uniformity of a bent function would be 4 but such bent function cannot exist. No such differentially 6-uniform $(n, n - 2)$-function has been found so far and the existence of such function, and more generally of a differentially δ-uniform $(n, n-k)$-function, $k \geq 2, n > 2k$, with $\delta < 2^{k+1}$ is an open problem (differentially 2^{k+1}-uniform $(n, n - k)$-functions are easily found by composing on the left any APN function by a surjective affine $(n, n - k)$-function).

3.11 Is It Possible, Given the Set of Those Pairs $(a, b) \in \mathbb{F}_2^n \times \mathbb{F}_2^m$ Such That, for Some Hidden APN Function F, the Equation $F(x) + F(x + a) = b$ Has Solutions, to Reconstruct F or Any Other Function Admitting the Same Associated Set of Pairs?

It is observed in [12] that, given an (n, n)-function F, the function $\gamma_F(a, b)$ taking value 1 if $a \neq 0$ and if the equation $F(x) + F(x + a) = b$ admits solutions, and value 0 otherwise, has weight $2^{2n-1} - 2^{n-1}$ if and only if F is APN. It is proved in [12] that F is AB if and only if γ_F is bent. A possible way for finding more APN or AB functions is to search for those vectorial functions whose associated function γ_F equals that of an already known APN function.

– Is it possible to find a systematic way, given an APN function F, to build another function F' such that $\gamma_{F'} = \gamma_F$?

4 More Open Problems

4.1 Use the Numerical Normal Form to Build Nonlinear Boolean Functions (such as Bent Functions) from Other Functions, Possibly Linear

The *Numerical Normal Form* (NNF) is a representation of n-variable Boolean functions as elements of $\mathbb{Z}[x_1, \cdots, x_n]/(x_1^2 - x_1, \cdots, x_n^2 - x_n)$ [7,13,14]. The NNF $f(x) = \sum_{I \subseteq \{1,\ldots,n\}} \lambda_I \prod_{i \in I} x_i$ of every function f from $\mathbb{F}_2^n$ to $\mathbb{Z}$ (in particular, of every n-variable Boolean function, considered as valued in $\{0,1\}$) exists and is unique. We have: $f(x) = \sum_{I \subseteq supp(x)} \lambda_I$. The reverse formula is:

$$\forall I \subseteq \{1,\ldots,n\}, \lambda_I = (-1)^{|I|} \sum_{x \in \mathbb{F}_2^n \,|\, supp(x) \subseteq I} (-1)^{w_H(x)} f(x), \qquad (6)$$

where w_H denotes the Hamming weight. The ANF of any Boolean function can be deduced from its NNF by reducing it modulo 2. Conversely, the NNF can be deduced from the ANF as follows. We have:

$$f(x) = \sum_{I \subseteq \{1,\dots,n\}} a_I \prod_{i \in I} x_i \,[\text{mod } 2] \iff (-1)^{f(x)} = \prod_{I \subseteq \{1,\dots,n\}} (-1)^{a_I \, \Pi_{i \in I} \, x_i}$$

$$\iff 1 - 2\, f(x) = \prod_{I \subseteq \{1,\dots,n\}} (1 - 2\, a_I \prod_{i \in I} x_i).$$

Expanding this last equality gives the NNF of $f(x)$ and we have [13]:

$$\lambda_I = \sum_{k=1}^{2^n} (-2)^{k-1} \sum_{\substack{\{I_1,\dots,I_k\}\,| \\ I_1 \cup \dots \cup I_k = I}} a_{I_1} \cdots a_{I_k}, \tag{7}$$

where "$\{I_1, \dots, I_k\}$" means that the multi-indices $I_1, \dots, I_k$ are all distinct, and that each such set appears only once.

We call the degree of the NNF of a function its *numerical degree*. Since the ANF is the mod 2 version of the NNF, the numerical degree is always bounded below by the algebraic degree. We have also, for every $a \neq 0$, that $W_f(a) = (-1)^{w_H(a)+1} \sum_{supp(a) \subseteq I \subseteq \{1,\dots,n\}} 2^{n+1-|I|} \lambda_I$. Hence, for every $a \in \mathbb{F}_2^n$ of Hamming weight strictly larger than the numerical degree, the Walsh transform at a is null.

We introduce now a transformation on Boolean functions deduced from the NNF: given a Boolean function f expressed by its ANF:

$$f(x) = \sum_{I \subseteq \{1,\dots,n\}} a_I \prod_{i \in I} x_i \,[\text{mod } 2],$$

we consider its NNF. Every coefficient in its NNF can be expanded in binary representation $\lambda_I = \sum_{j \geq 0} \lambda_{I,j} 2^j$. We denote by $f^{(j)}$ the following function expressed by its ANF:

$$f^{(j)}(x) = \sum_{I \subseteq \{1,\dots,n\}} \lambda_{I,j} \prod_{i \in I} x_i \,[\text{mod } 2].$$

We have of course $f = f^{(0)}$.

For example, according to Relation (7), we have:

$$f^{(1)}(x)) = \sum_{I \subseteq \{1,\dots,n\}} \left(\sum_{\substack{\{I_1,I_2\}\,| \\ I_1 \cup I_2 = I}} a_{I_1} a_{I_2} \right) \prod_{i \in I} x_i \,[\text{mod } 2],$$

that is

$$f^{(1)}(x) = \left(\frac{1}{2} \left[\left(\sum_{I \subseteq \{1,\dots,n\}} a_I \prod_{i \in I} x_i \right)^2 - \sum_{I \subseteq \{1,\dots,n\}} a_I \prod_{i \in I} x_i \right] \right) \,[\text{mod } 2]. \tag{8}$$

Clearly, the mapping $f \mapsto f^{(1)}$ is neither $\mathbb{F}_2$-linear nor bijective (all coordinate functions have the same image 0 by this mapping). It is quadratic in the sense that $(f, g) \mapsto f^{(1)} + g^{(1)} + (f + g)^{(1)}$ [mod 2] is $\mathbb{F}_2$-bilinear. Its behavior with respect to $\mathbb{F}_2$-affine equivalence is chaotic: if $f(x) = x_1$, then $f^{(1)}(x) = 0$, and if $f(x)$ is obtained from x_1 by composing it by $L(x) = (\sum_{i=1}^{n} x_i$ [mod 2]$, x_2, \cdots, x_n)$, the resulting function $f(x) = \sum_{i=1}^{n} x_i$ [mod 2] gives $f^{(1)}(x) = \sum_{1 \le i < j \le n} x_i x_j$ [mod 2], which is bent.

Note that we can express $f^{(1)}(x)$ by means of the values of f: we know (for instance by reducing Relation (6) modulo 2) that, for every $I \subseteq \{1, \ldots, n\}$, $a_I = \sum_{x \in \mathbb{F}_2^n \mid supp(x) \subseteq I} f(x)$ [mod 2]. Hence $1 - 2a_I = \prod_{x \in \mathbb{F}_2^n \mid supp(x) \subseteq I} (1 - 2f(x))$ and $a_I = \sum_{k=1}^{2^{|I|}} (-2)^{k-1} \sum_{\substack{\{x^{(1)}, \ldots, x^{(k)}\} \mid x^{(1)}, \ldots, x^{(k)} \in \mathbb{F}_2^n \\ supp(x^{(1)}) \subseteq I, \cdots, supp(x^{(k)}) \subseteq I}} f(x^{(1)}) \cdots f(x^{(k)})$. Note that this gives the coefficient of $\prod_{i \in I} x_i$ in the ANF of $f^{(1)}$, since this coefficient equals by definition $\frac{\lambda_I - a_I}{2}$ [mod 2]. This also implies that $\sum_{I \subseteq \{1, \ldots, n\}} a_I \prod_{i \in I} x_i =$

$$\sum_{I \subseteq supp(x)} a_I = \sum_{k=1}^{2^{w_H(x)}} (-2)^{k-1} \sum_{I \subseteq supp(x)} \sum_{\substack{\{x^{(1)}, \ldots, x^{(k)}\} \mid x^{(1)}, \ldots, x^{(k)} \in \mathbb{F}_2^n \\ supp(x^{(1)}) \subseteq I, \cdots, supp(x^{(k)}) \subseteq I}} f(x^{(1)}) \cdots f(x^{(k)}),$$

and therefore, according to Relation (8) and using that for every integers λ, μ, we have $(\frac{1}{2}[(\lambda + 2\mu)^2 - (\lambda + 2\mu)])$ [mod 2] $= (\frac{1}{2}[\lambda^2 - \lambda] + \mu)])$ [mod 2]:

$$f^{(1)}(x) = \left(\frac{1}{2} \left[\left(\sum_{I \subseteq supp(x)} \sum_{\substack{x' \in \mathbb{F}_2^n \mid \\ supp(x') \subseteq I}} f(x') \right)^2 - \sum_{I \subseteq supp(x)} \sum_{\substack{x' \in \mathbb{F}_2^n \mid \\ supp(x') \subseteq I}} f(x') \right] \right.$$

$$\left. + \sum_{I \subseteq supp(x)} \sum_{\substack{\{x', x''\} \mid x', x'' \in \mathbb{F}_2^n \\ supp(x') \subseteq I, supp(x'') \subseteq I}} f(x') f(x'') \right) \text{ [mod 2]},$$

that is: $f^{(1)}(x) =$

$$\left(\frac{1}{2} \left[\left(\sum_{\substack{x' \in \mathbb{F}_2^n \mid \\ supp(x') \subseteq supp(x)}} 2^{w_H(x+x')} f(x') \right)^2 - \sum_{\substack{x' \in \mathbb{F}_2^n \mid \\ supp(x') \subseteq supp(x)}} 2^{w_H(x+x')} f(x') \right] \right.$$

$$\left. + \sum_{\substack{\{x', x''\} \mid x', x'' \in \mathbb{F}_2^n \\ supp(x') \subseteq supp(x), supp(x'') \subseteq supp(x)}} 2^{w_H(x) - w_H(x' \vee x'')} f(x') f(x'') \right) \text{ [mod 2]} =$$

$$\left(\sum_{\substack{x' \in \mathbb{F}_2^n \mid w_H(x') = w_H(x) - 1; \\ supp(x') \subseteq supp(x)}} f(x') + \sum_{\substack{\{x', x''\} \mid x', x'' \in \mathbb{F}_2^n \\ x' \vee x'' = x}} f(x') f(x'') \right) \text{ [mod 2]},$$

where $x' \vee x''$ is defined by $supp(x' \vee x'') = supp(x') \cup supp(x'')$.

The following questions have not yet been studied:

1. Find, for $j > 1$, the expression of $f^{(j)}(x)$ by means of the values of f;
2. Find, for $j > 1$, the expression of the ANF of $f^{(j)}$ by means of the ANF of f;
3. Given a function f, determine all the functions $f'^{(1)}$ where f' is affinely equivalent to f; same question for $f'^{(j)}$;
4. Determine the set of $f^{(1)}$ when f ranges over the set of all n-variable Boolean functions (resp. the set of all n-variable Boolean functions of algebraic degree at most k); same questions for $f^{(j)}$;
5. Find a necessary and sufficient condition on f such that $f^{(1)}$ is bent; idem for $f^{(j)}$;
6. Find an expression of the Walsh transform of $f^{(1)}$ by means of the Walsh transform of f;
7. Find expressions by means of f of the cryptographic parameters of $f^{(1)}$ such as the resiliency order, the nonlinearity, the algebraic immunity or the fast algebraic immunity;
8. Study other transformations of a similar kind, related to the NNF.

4.2   About the Relation Between the Nonlinearity of a Boolean Function and the Difference Distribution of Its Support

It is well-known that a Boolean function f is bent (i.e. achieves optimal nonlinearity $2^{n-1} - 2^{n/2-1}$) if and only if its support $S_f = \{x \in \mathbb{F}_2^n ;\ f(x) = 1\}$ is a non-trivial difference set, that is, S_f is such that the size of the set $\{(x, y) \in S_f^2;\ x + y = z\}$ (whose list of values can be called the difference distribution of S_f) is independent of $z \neq 0$ in $\mathbb{F}_2^n$, and S_f has size different from $0, 1, 2^n - 1$ and 2^n (see [20]). A natural question is then:

– Is there a more general relationship between the nonlinearity of a Boolean function and the difference distribution of its support?

Given an n-variable Boolean function f, let

$$M_f = \frac{\max_{z \in (\mathbb{F}_2^n)^*} \#\left\{(x, y) \in S_f^2;\ x + y = z\right\} + \min_{z \in (\mathbb{F}_2^n)^*} \#\left\{(x, y) \in S_f^2;\ x + y = z\right\}}{2}$$

and

$$E_f = \frac{\max_{z \in (\mathbb{F}_2^n)^*} \#\left\{(x, y) \in S_f^2;\ x + y = z\right\} - \min_{z \in (\mathbb{F}_2^n)^*} \#\left\{(x, y) \in S_f^2;\ x + y = z\right\}}{2},$$

where $(\mathbb{F}_2^n)^*$ denotes $\mathbb{F}_2^n \setminus \{0\}$ and $\#$ denotes the cardinality (we use this notation here instead of $|\dots|$ to avoid confusion with absolute value).

Then we have that, for every $z \neq 0$: $\left| \# \left\{ (x,y) \in S_f^2;\ x + y = z \right\} - M_f \right| \leq E_f$ and:

$$\left| \sum_{(x,y) \in S_f^2, x \neq y} (-1)^{(x+y) \cdot a} - M_f \sum_{z \in (\mathbb{F}_2^n)^*} (-1)^{z \cdot a} \right| =$$

$$\left| \sum_{z \in (\mathbb{F}_2^n)^*} \left(\# \left\{ (x,y) \in S_f^2;\ x + y = z \right\} - M_f \right) (-1)^{z \cdot a} \right| \leq$$

$$\sum_{z \in (\mathbb{F}_2^n)^*} \left| \# \left\{ (x,y) \in S_f^2;\ x + y = z \right\} - M_f \right| \leq (2^n - 1) E_f.$$

This implies for every $a \in (\mathbb{F}_2^n)^*$, since $\sum_{z \in (\mathbb{F}_2^n)^*} (-1)^{z \cdot a} = -1$, that the square of the Fourier transform of f at a: $(\widehat{f})^2(a) = \sum_{(x,y) \in S_f^2} (-1)^{(x+y) \cdot a} = \# S_f + \sum_{(x,y) \in S_f^2, x \neq y} (-1)^{(x+y) \cdot a}$ satisfies:

$$\left| (\widehat{f})^2(a) - \# S_f + M_f \right| = \left| \sum_{(x,y) \in S_f^2, x \neq y} (-1)^{(x+y) \cdot a} - M_f \sum_{z \in (\mathbb{F}_2^n)^*} (-1)^{z \cdot a} \right|$$

$$\leq (2^n - 1) E_f.$$

We deduce the following result, using Relations (2), (4) and (5):

Proposition 2. *Let f be any n-variable Boolean function. Let S_f, M_f and E_f be defined as above. Then:*

$$nl(f) \geq$$

$$2^{n-1} - \max \left(|2^{n-1} - \# S_f|, \sqrt{\# S_f - M_f + (2^n - 1) E_f} \right). \qquad (9)$$

Note that $E_f = 0$ (and therefore, $M_f = \frac{\# S_f (\# S_f - 1)}{2^n - 1}$) if and only if S_f is a difference set. Then (9) is an equality since $\# S_f \in \{0, 1\}$ implies $M_f = 0$ and the equality is easily verified, $\# S_f \in \{2^n - 1, 2^n\}$ implies equality as well by considering the complement of S_f, and $\# S_f = 2^{n-1} \pm 2^{n/2-1}$ (which are the two possible weights of a bent function) implies $M_f = (2^{n-1} \pm 2^{n/2-1}) \frac{2^{n-1} \pm 2^{n/2-1} - 1}{2^n - 1} = \frac{2^{2n-2} \pm 2^{3n/2-1} - 2^{n-2} - \pm 2^{n/2-1}}{2^n - 1} = 2^{n-2} \pm 2^{n/2-1}$ and $\# S_f - M_f = 2^{n-2}$. Hence:

Proposition 3. *Any bent function achieves (9) with equality.*

According to the observations above, (9) is an equality if and only if either f has large or small Hamming weight (which are cases of little interest), or there exists $a \neq 0$ such that, for every $z \neq 0$, we have $\# \left\{ (x,y) \in S_f^2;\ x + y = z \right\} = M_f + (-1)^{z \cdot a} E_f$. We state then the following problem:

– Determine precisely all functions f such that (9) is an equality.

Application to the Open Problem of Bounding Below the Nonlinearity of the CF Function. Inequality (9) shows that if E_f is small (that is, if the support S_f of f is nearly a difference set) then f has good nonlinearity. Let us take the example of a function related to the discrete logarithm in $\mathbb{F}_{2^n}$, and which is sometimes called the CF function, because of ref. [6]. Its nonlinearity has been computed for $n \leq 38$ and the values obtained are good (see [36]) but it could not be mathematically proved, by the way of an efficient lower bound, that the nonlinearity is good for every n. This poses an open problem in cryptography, which happens to be also, in different terms, an open problem in sequence theory, see [11]. A series of papers have proved lower bounds on this nonlinearity. Each bound slightly improves upon the previous one, but all these bounds seem weak[3].

The support S_f of the CF function equals $\{\alpha^i, i = 0, \ldots, 2^{n-1} - 1\}$, up to EA-equivalence, where α is a primitive element of $\mathbb{F}_{2^n}$. We have a first question:

– Is it possible to bound E_f in the case of the CF function and to deduce a sharper bound on the nonlinearity of the CF function by using Inequality (9)?

It is possible to study more in details the Fourier transform of the CF function. To this aim, let us denote for every $s \leq 2^n - 2$ by f_s the Boolean function of support $\{\alpha^i, i = 0, \ldots, s\}$. Let us also denote by I_s the set $\{1 + \alpha^k, k = 1, \ldots, s\}$ and by g_s its indicator. Note that $g_s(x) = f_s(x+1) + \delta_0(x)$ [mod 2], where δ_0 is the Dirac (Kronecker) symbol, and therefore $f_s(x) = g_s(x+1) + \delta_1(x)$ [mod 2]. These equalities imply that the Fourier transform $\widehat{g_s}(a) = \sum_{x \in \mathbb{F}_2^n} g_s(x)(-1)^{tr_n(ax)}$ equals $(-1)^{tr_n(a)} \widehat{f_s}(a) - 1$ and bounding the nonlinearity of the CF function can be then deduced from the study of the Fourier transform of g_s for $s = 2^{n-1} - 1$.

We have $g_s \otimes g_s(z) = \#\{(i,j); \alpha^i + \alpha^j = z, 1 \leq i,j \leq s\} = \#\{(i,j); \alpha^{i+1} + \alpha^{j+1} = z, 1 \leq i,j \leq s-1\} + 2\#\{j; 1 + \alpha^j = \alpha^{-1}z, 1 \leq j \leq s-1\} + \delta_0(z) = g_{s-1} \otimes g_{s-1}(\alpha^{-1}z) + 2g_{s-1}(\alpha^{-1}z) + \delta_0(z)$. By applying the Fourier transform, we deduce that, for every a:

$$\left(\widehat{g_{s-1}}(\alpha a) + 1\right)^2 = \widehat{g_s}^2(a)$$

that is

$$|\widehat{g_{s-1}}(\alpha a) + 1| = |\widehat{g_s}(a)|. \tag{10}$$

Hence, since g_1 has Hamming weight 1 and $\widehat{g_1}$ takes then its values in $\{-1, 1\}$, we have that $|\widehat{g_s}(a)|$ equals s minus twice the number of times that $\widehat{g_i}(\alpha^{s-i}a)$ is negative for $1 \leq i < s$. This poses the question:

– For every $a \neq 0$, bound below the number of $1 \leq i < 2^{n-1} - 1$ such that $\widehat{g_i}(\alpha^{-i}a) < 0$. Deduce an efficient lower bound on the nonlinearity of the CF function.

With a slightly different viewpoint, we have also:

---

[3] Unless the nonlinearity of the CF function becomes much worse for values of n which are too large for allowing its computation.

Proposition 4. *For every $1 \leq s \leq 2^n - 2$, let f_s and g_s be the Boolean functions of supports $\{\alpha^i, i = 0, \ldots, s\}$ and $\{1 + \alpha^k, k = 1, \ldots, s\}$, respectively. Let $u_i^+ = \max\{\widehat{g_i}(a); \widehat{g_i}(a) > 0\}$; $u_i^- = \max\{|\widehat{g_i}(a)|; \widehat{g_i}(a) < 0\}$ and $u_i = \max(u_i^+, u_i^-)$. Then the nonlinearity of the CF function $f_{2^{n-1}-1}$ satisfies $nl(f_{2^{n-1}-1}) = 2^{n-1} - u_{2^{n-1}-1}$ and, for every $i \geq 2$, we have:*

1. *if $u_{i-1}^- \leq u_{i-1}^+$ then $u_i = u_{i-1}^+ + 1 = u_{i-1} + 1$;*
2. *if $u_{i-1}^- = u_{i-1}^+ + 1$ then $u_i = u_{i-1}^+ + 1 = u_{i-1}$;*
3. *if $u_{i-1}^- \geq u_{i-1}^+ + 2$ then $u_i = u_{i-1}^- - 1 = u_{i-1} - 1$.*

This poses the following question:

- Bound efficiently, above, the number of times each case 1-2 above occurs for $s \leq 2^{n-1} - 2$ and deduce an efficient lower bound on the nonlinearity of the CF function.

4.3   Find New Ways of Deducing Bent Vectorial Functions from Almost Bent (AB) Functions

As already recalled in Subsect. 3.11, it is proved in [12] that an (n, n)-function F (n odd) is AB if and only if the Boolean function $\gamma_F(a, b)$ equal to 1 if $a \neq 0$ and if the equation $F(x) + F(x + a) = b$ admits solutions, is bent. This has led to new bent Boolean functions, studied more in details in [3].

It is also possible to derive bent vectorial $(n, n/2)$-functions from APN (n, n)-functions (n even) by finding an $n/2$-dimensional subspace E of $\mathbb{F}_2^n$ such that any component function $v \cdot F$, $v \in E \setminus \{0\}$, is bent (see [8] where this is done with the Gold and Kasami APN functions).

We propose below a new way to hopefully deduce new bent vectorial functions from AB power functions. Let $F(x) = x^d$ be an APN power function on $\mathbb{F}_{2^n}$, where n is any positive integer. We know that, since the function $F(x) + F(x+1)$ is invariant under the translation $x \mapsto x + 1$, there exists a function I on the hyperplane $H = \{y \in \mathbb{F}_{2^n}; tr_n(y) = 0\}$ such that $F(x) + F(x+1) = I(x + x^2)$, and since $F(x) + F(x + 1)$ is 2-to-1, then I is injective on H. For every $a \neq 0$ in $\mathbb{F}_{2^n}$, we have $F(x) + F(x + a) = a^d I\left(\left(\frac{x}{a}\right) + \left(\frac{x}{a}\right)^2\right)$.

Any permutation whose restriction to H equals I is such that $F(x) + F(x + a) = a^d P\left(\left(\frac{x}{a}\right) + \left(\frac{x}{a}\right)^2\right)$. The number of these permutations P equals $2^{n-1}!$

For every $u, v \in \mathbb{F}_{2^n}$, the squared value of $W_{v \cdot F}(u)$ equals:

$$\left(\sum_{x \in \mathbb{F}_{2^n}} (-1)^{tr_n(vF(x) + ux)}\right)^2 = \sum_{x, y \in \mathbb{F}_{2^n}} (-1)^{tr_n(v(F(x) + F(y)) + u(x+y))}$$

$$= \sum_{x, a \in \mathbb{F}_{2^n}} (-1)^{tr_n(v(F(x) + F(x+a)) + ua)}$$

$$= 2^n + \sum_{\substack{x,a\in\mathbb{F}_{2^n} \\ a\neq 0}} (-1)^{tr_n\left(v\,a^d\,P\left((\frac{x}{a})+(\frac{x}{a})^2\right)+ua\right)}$$

$$= 2^n + 2 \sum_{\substack{y,a\in\mathbb{F}_{2^n} \\ tr_n(y)=0;\,a\neq 0}} (-1)^{tr_n\left(v\,a^d\,P(y)+ua\right)}.$$

We deduce that

$$\left(\sum_{x\in\mathbb{F}_{2^n}} (-1)^{tr_n(vF(x)+ux)}\right)^2 = 2^n + \sum_{\substack{y,a\in\mathbb{F}_{2^n} \\ a\neq 0}} (-1)^{tr_n\left(v\,a^d\,P(y)+ua\right)}$$

$$+ \sum_{\substack{y,a\in\mathbb{F}_{2^n} \\ a\neq 0}} (-1)^{tr_n\left(v\,a^d\,P(y)+y+ua\right)}.$$

Since P is a permutation, we have $\sum_{y\in\mathbb{F}_{2^n}} (-1)^{tr_n(v\,a^d\,P(y)+ua)} = 0$, for every $v \neq 0$, every u and every $a \neq 0$ in $\mathbb{F}_{2^n}$. Hence, since for $a = 0$ we have $\sum_{y\in\mathbb{F}_{2^n}} (-1)^{tr_n(v\,a^d\,P(y)+y+ua)} = \sum_{y\in\mathbb{F}_{2^n}} (-1)^{tr_n(y)} = 0$, we deduce that for every $v \neq 0$ and every u in $\mathbb{F}_{2^n}$:

$$\left(\sum_{x\in\mathbb{F}_{2^n}} (-1)^{tr_n(vF(x)+ux)}\right)^2 = 2^n + \sum_{y,a\in\mathbb{F}_{2^n}} (-1)^{tr_n\left(v\,a^d\,P(y)+y+ua\right)}.$$

If F is AB (n odd), then we have $\left(\sum_{x\in\mathbb{F}_{2^n}} (-1)^{tr_n(vF(x)+ux)}\right)^2 \in \{0, 2^{n+1}\}$ for every $v \neq 0$ and every u in $\mathbb{F}_{2^n}$ and we deduce that

$$\sum_{y,a\in\mathbb{F}_{2^n}} (-1)^{tr_n\left(v\,a^d\,P(y)+y+ua\right)} = \pm 2^n, \forall v \in \mathbb{F}_{2^n}, v \neq 0, \forall u \in \mathbb{F}_{2^n}. \qquad (11)$$

Note that the function $(a, y) \to a^d\,P(y)$ is bent if and only if

$$\sum_{y,a\in\mathbb{F}_{2^n}} (-1)^{tr_n\left(v\,a^d\,P(y)+wy+ua\right)} = \pm 2^n, \forall v \in \mathbb{F}_{2^n}, v \neq 0, \forall u, w \in \mathbb{F}_{2^n},$$

and (11) is this equality when $w = 1$. Note also that for $w = 0$, the equality is satisfied since we have $\sum_{y,a\in\mathbb{F}_{2^n}} (-1)^{tr_n(v\,a^d\,P(y)+ua)} = 2^n$. This leads to the question:

– For the known power AB functions, can we choose P such that the function $(a, y) \to a^d\,P(y)$ is bent?

Remarks

1. If F is a Gold AB function (that is, $F(x) = x^{2^k+1}$, where $(n, k) = 1$ and n is odd) with k odd, then the answer is yes, since we can then take $P(y) = 1 + y + y^2 + y^4 + \cdots + y^{2^{k-1}}$ and the function $(a, y) \to a^{2^k+1}\,P(y)$ is then

linearly equivalent to a Maiorana-McFarland bent function (see [7]), since $a \to a^{2^k+1}$ is bijective and $y \to y + y^2 + y^4 + \cdots + y^{2^{k-1}}$ is linear bijective (indeed, $y + y^2 + y^4 + \cdots + y^{2^{k-1}} = 0$ implies $y + y^{2^k} = 0$ which implies $y \in \mathbb{F}_2$ and $y = 1$ gives $y + y^2 + y^4 + \cdots + y^{2^{k-1}} = 1$).

2. If F is a Gold AB function with k is even, then the answer is yes too, since we can then take $P(y) = 1 + y + y^2 + y^4 + \cdots + y^{2^{k-1}} + tr_n(y)$ and the function $(a, y) \to a^{2^k+1} P(y)$ is then linearly equivalent to a Maiorana-McFarland bent function as well, since $y \to y + y^2 + y^4 + \cdots + y^{2^{k-1}} + tr_n(y)$ is linear bijective (indeed, $y + y^2 + y^4 + \cdots + y^{2^{k-1}} + tr_n(y) = 0$ implies $y + y^{2^k} = 0$ which implies $y \in \mathbb{F}_2$ and $y = 1$ gives $y + y^2 + y^4 + \cdots + y^{2^{k-1}} + tr_n(y) = 1$).

In both cases 1 and 2 above, the bent function is not new, but other functions P could be investigated and this leads to the sub-question:

– Is it possible to find nonlinear permutations P of $\mathbb{F}_{2^n}$ such that $(a, y) \in \mathbb{F}_{2^n} \times \mathbb{F}_{2^n} \to a^{2^k+1} P(y)$ is a new bent function?

4.4 Find a Direct Proof of APN-ness of the Kasami Functions in Even Dimension Which Would Use the Relationship Between These Functions and the Gold Functions

The proof by Hans Dobbertin in [21] of the fact that Kasami functions $F(x) = x^{2^{2k}-2^k+1}$, where $(i, n) = 1$, are AB (and therefore APN) for n odd uses that these functions are the (commutative) composition of a Gold function and of the inverse of another Gold function. This proof is particularly simple. The direct proof in [22] that the Kasami functions above are APN for n even is harder, as well as the determination in [23, Theorem 11] of their Walsh spectrum, which also allows to prove their APNness, and which uses a similar but slightly more complex relation to the Gold functions when n is not divisible by 6. It would be interesting to see if, for n odd and for n even, these relations between the Kasami functions and the Gold functions can lead to alternative direct proofs, hopefully simpler, of the APN-ness of Kasami functions.

Since the Kasami function is a power function, it is APN if and only if, for every $b \in \mathbb{F}_{2^n}$ the system

$$\begin{cases} X + Y & = 1 \\ F(X) + F(Y) = b \end{cases} \tag{12}$$

has at most one pair $\{X, Y\}$ of solutions in $\mathbb{F}_{2^n}$.

We first address the case n odd. Then, $2^k + 1$ is coprime with $2^n - 1$ and $F(x) = G_2 \circ G_1^{-1}(x)$, where $G_1(x)$ and $G_2(x)$ are respectively the Gold functions x^{2^k+1} and $x^{2^{3k}+1}$. Hence, F is APN if and only if the system

$$\begin{cases} x^{2^k+1} + y^{2^k+1} & = 1 \\ x^{2^{3k}+1} + y^{2^{3k}+1} = b \end{cases} \tag{13}$$

has at most one pair $\{x, y\}$ of solutions.

Let $y = x + u$. Then $u \neq 0$. The system (13) is equivalent to:

$$\begin{cases} \left(\frac{x}{u}\right)^{2^k} + \left(\frac{x}{u}\right) = \frac{1}{u^{2^k+1}} + 1 \\ \left(\frac{x}{u}\right)^{2^{3k}} + \left(\frac{x}{u}\right) = \frac{b}{u^{2^{3k}+1}} + 1 \end{cases}$$

or equivalently $\begin{cases} \left(\frac{x}{u}\right)^{2^k} + \left(\frac{x}{u}\right) = \frac{1}{u^{2^k+1}} + 1 \\ \frac{1}{u^{2^k+1}} + 1 + \left(\frac{1}{u^{2^k+1}} + 1\right)^{2^k} + \left(\frac{1}{u^{2^k+1}} + 1\right)^{2^{2k}} = \frac{b}{u^{2^{3k}+1}} + 1 \end{cases}$

that is, by simplifying and multiplying the second equation by $u^{2^{3k}+2^{2k}}$:

$$\begin{cases} \left(\frac{x}{u}\right)^{2^k} + \left(\frac{x}{u}\right) = \frac{1}{u^{2^k+1}} + 1 \\ u^{2^{3k}+2^{2k}-2^k-1} + u^{2^{3k}-2^k} + 1 = b\,u^{2^{2k}-1} \end{cases}$$

that is, denoting $v = u^{2^{2k}-1}$ and $c = b + 1$:

$$\begin{cases} \left(\frac{x}{u}\right)^{2^k} + \left(\frac{x}{u}\right) = \frac{1}{u^{2^k+1}} + 1 \\ (v+1)^{2^k+1} + cv = 0 \end{cases}$$

Proving that F is APN is equivalent to proving that, for every $c \in \mathbb{F}_{2^n}$, the second equation can be satisfied by at most one value of v such that the first equation can admit solutions, i.e. such that $tr_n\left(\frac{1}{u^{2^k+1}} + 1\right) = 0$. Reference [28] studies the equation $x^{2^k+1} + c(x+1) = 0$. Unfortunately, the results in this paper do not seem to give directly the end of this proof.

– For n odd, is it possible to complete this proof?

Let us now consider the case n even. Note first that system (12) has a solution such that $X = 0$ or $Y = 0$ if and only if $b = 1$. We restrict now ourselves to the case where n is not divisible by 6. Then $(\frac{2^n-1}{3}, 3) = 1$ and every element X of $\mathbb{F}_{2^n}^*$ can be written (in 3 different ways) in the form wx^{2^k+1}, $w \in \mathbb{F}_4^*, x \in \mathbb{F}_{2^n}^*$. Indeed, the function $x \mapsto x^{2^k+1}$ is 3-to-1 from $\mathbb{F}_{2^n}^*$ to the set of cubes of $\mathbb{F}_{2^n}^*$, and every integer i being, by the Bézout theorem, the linear combination over $\mathbb{Z}$ of $\frac{2^n-1}{3}$ and 3, every element α^i of $\mathbb{F}_{2^n}^*$ (where α is primitive) is the product of a power of $\alpha^{\frac{2^n-1}{3}}$ and of a power of α^3. So, F is APN if and only if the system

$$\begin{cases} wx^{2^k+1} + w'y^{2^k+1} = 1 \\ x^{2^{3k}+1} + y^{2^{3k}+1} = b \end{cases} \tag{14}$$

(since $2^{2k} - 2^k + 1 = (2^k + 1)^2 - 3 \cdot 2^k$ is divisible by 3), where $w, w' \in \mathbb{F}_4^*$ and $x, y \in \mathbb{F}_{2^n}^*$, has no solution for $b = 1$ and has at most one pair $\{wx^{2^k+1}, w'y^{2^k+1}\}$ of solutions for every $b \neq 1$.

We consider the case $x^{2^k+1} = y^{2^k+1}$ (and $w \neq w'$) apart. In this case, the first equation $(w + w')x^{2^k+1} = 1$ is equivalent to $x \in \mathbb{F}_4^*$ and $w + w' = 1$.

Then because of the second equation, we have $b = 0$. Hence, for $b = 0$, we have two solutions such that $x^{2^k+1} = y^{2^k+1}$ (since w and w' are nonzero) and for $b \neq 0$ we have none. Hence, F is APN if and only if the system

$$\begin{cases} wx^{2^k+1} + w'y^{2^k+1} = 1 \\ x^{2^{3k}+1} + y^{2^{3k}+1} = b \end{cases} \tag{15}$$

where $w, w' \in \mathbb{F}_4^*$ and $x, y \in \mathbb{F}_{2^n}$ are such that $x^{2^k+1} \neq y^{2^k+1}$, has no solution for $b \in \mathbb{F}_2$ and has at most one pair $\{wx^{2^k+1}, w'y^{2^k+1}\}$ of solutions for every $b \notin \mathbb{F}_2$. Since $x^{2^k+1} \neq y^{2^k+1}$, we can as above denote $y = x + u$ where $u \neq 0$, $v = u^{2^{2k}-1}$ and $c = b + 1$, and we obtain the system:

$$\begin{cases} (w + w')\left(\frac{x}{u}\right)^{2^k+1} + w'\left(\frac{x}{u}\right)^{2^k} + w'\left(\frac{x}{u}\right) = \frac{1}{u^{2^k+1}} + w' \\ (v+1)^{2^k+1} + cv = 0 \end{cases}$$

where $u \neq 0$, $v \neq 0$. Reference [28] can be useful here also, but does not suffice.

- For n even not divisible by 6, is it possible to complete this proof?
- For n divisible by 6, is it possible to adapt the method?

Acknowledgement. We wish to thank Lilya Budaghyan, Faruk Gologlu and Sihem Mesnager for useful information.

References

1. Browning, K., Dillon, J., McQuistan, M.: APN polynomials and related codes. Special volume of Journal of Combinatorics, Information and System Sciences, honoring the 75-th birthday of Prof. D.K.Ray-Chaudhuri 34, 135–159 (2009)
2. Browning, K., Dillon, J., McQuistan, M., Wolfe, A.: An APN permutation in dimension six. Contemp. Math. **58**, 33–42 (2010)
3. Budaghyan, L., Carlet, C., Helleseth, T.: On bent functions associated to AB functions. In: Proceedings of IEEE Information Theory Workshop (2011)
4. Budaghyan, L., Carlet, C., Pott, A.: New classes of almost bent and almost perfect nonlinear functions. IEEE Trans. Inform. Theory **52**(3), 1141–1152 (2006)
5. Carlet, C.: Generalized partial spreads. IEEE Trans. Inform. Theory **41**(5), 1482–1487 (1995)
6. Carlet, C., Feng, K.: An infinite class of balanced functions with optimal algebraic immunity, good immunity to fast algebraic attacks and good nonlinearity. In: Pieprzyk, J. (ed.) ASIACRYPT 2008. LNCS, vol. 5350, pp. 425–440. Springer, Heidelberg (2008)
7. Carlet, C.: Boolean functions for cryptography and error correcting codes. In: Crama, Y., Hammer, P.L. (eds.) Boolean Models and Methods in Mathematics, Computer Science, and Engineering, pp. 257–397. Cambridge University Press, Cambridge (2010)
8. Carlet, C.: Vectorial Boolean Functions for Cryptography, Idem, pp. 398–469 (2010)

9. Carlet, C.: Relating three nonlinearity parameters of vectorial functions and building APN functions from bent. Des. Codes Crypt. **59**(1), 89–109 (2011)

10. Carlet, C.: Open problems on binary bent functions. In: Proceedings of the Conference Open Problems in Mathematical and Computational Sciences, September 18–20, 2013, in Istanbul, Turkey, pp. 203–241. Springer (2014)

11. Carlet. C.: A survey on nonlinear boolean functions with optimal algebraic immunity suitable for stream ciphers. In: Proceedings of the SMF-VMS Conference, Hué, Vietnam, 20–24 August 2012. (Special issue of the Vietnam Journal of Mathematics, Volume 41, Issue 4, Page 527–541, 2013)

12. Carlet, C., Charpin, P., Zinoviev, V.: Codes, bent functions and permutations suitable for DES-like cryptosystems. Des. Codes Crypt. **15**(2), 125–156 (1998)

13. Carlet, C., Guillot, P.: A new representation of boolean functions. In: Fossorier, M.P.C., Imai, H., Lin, S., Poli, A. (eds.) AAECC 1999. LNCS, vol. 1719, pp. 94–103. Springer, Heidelberg (1999)

14. Carlet, C., Guillot, P.: Bent, resilient functions and the numerical normal form. DIMACS Ser. Discrete Math. Theoret. Comput. Sci. **56**, 87–96 (2001)

15. Carlet, C., Klapper, A.: Upper bounds on the numbers of resilient functions and of bent functions. This paper was meant to appear in an issue of Lecture Notes in Computer Sciences dedicated to Philippe Delsarte, Editor Jean-Jacques Quisquater. But this issue finally never appeared. A shorter version has appeared in the Proceedings of the 23rd Symposium on Information Theory in the Benelux, Louvain-La-Neuve, Belgium (2002)

16. Carlet, C., Tang, D.: Enhanced Boolean functions suitable for the filter model of pseudo-random generator. Designs, Codes and Cryptography (to appear)

17. Courtois, N.T.: Fast algebraic attacks on stream ciphers with linear feedback. In: Boneh, D. (ed.) CRYPTO 2003. LNCS, vol. 2729, pp. 176–194. Springer, Heidelberg (2003)

18. Chepyzhov, V., Smeets, B.J.M.: On a fast correlation attack on certain stream ciphers. In: Davies, D.W. (ed.) EUROCRYPT 1991. LNCS, vol. 547, pp. 176–185. Springer, Heidelberg (1991)

19. Dillon, J.: A survey of bent functions. NSA Tech. J., 191–215 (1972). Special Issue

20. Dillon, J.F.: Elementary Hadamard Difference sets. Ph. D. Thesis, Univ. of Maryland (1974)

21. Dobbertin, H.: Another proof of Kasami's Theorem. Des. Codes Crypt. **17**, 177–180 (1999)

22. Dobbertin, H.: Kasami power functions, permutation polynomials and cyclic difference sets. In: Proceedings of the NATO-A.S.I. Workshop "Difference sets, sequences and their correlation properties", Bad Windsheim, Kluwer Verlag, pp. 133–158 (1998)

23. Dillon, J.F., Dobbertin, H.: New cyclic difference sets with Singer parameters. Finite Fields Appl. **10**, 342–389 (2004)

24. Edel, Y., Pott, A.: A new almost perfect nonlinear function which is not quadratic. Adv. Math. Commun. **3**(1), 59–81 (2009)

25. Gologlu, F.: Projective polynomials and their applications in cryptography. In: International Workshop on Boolean Functions and Their Applications, Bergen, September 2014. http://www.people.uib.no/lbu061/gologlu.pdf

26. Guillot, P.: Completed GPS covers all bent functions. J. Comb. Theory Ser. A **93**, 242–260 (2001)

27. Ding, C., Shan, W., Xiao, G. (eds.): The Stability Theory of Stream Ciphers. LNCS, vol. 561. Springer, Heidelberg (1991)

28. Helleseth, T.: Kholosha, Alexander: $x^{2^l+1} + x + a$ and related affine polynomials over $GF(2^k)$. Crypt. Commun. **2**(1), 85–109 (2010)
29. Langevin, P., Leander, G.: Counting all bent functions in dimension eight 99270589265934370305785861242880. Des. Codes Crypt. **59**(1–3), 193–205 (2011)
30. Liu, M., Lin, D., Pei, D.: Fast algebraic attacks and decomposition of symmetric boolean functions. IEEE Trans. Inform. Theory **57**, 4817–4821 (2011). A preliminary version of this paper was presented in ArXiv: 0910.4632v1 [cs.CR]. http://arxiv.org/abs/0910.4632
31. Massey, J.L.: Shift-register analysis and BCH decoding. IEEE Trans. Inf. Theory **15**, 122–127 (1969)
32. Nyberg, K.: Perfect nonlinear S-boxes. In: Davies, D.W. (ed.) EUROCRYPT 1991. LNCS, vol. 547, pp. 378–386. Springer, Heidelberg (1991)
33. Piret, G., Roche, T., Carlet, C.: PICARO – a block cipher allowing efficient higher-order side-channel resistance. In: Bao, F., Samarati, P., Zhou, J. (eds.) ACNS 2012. LNCS, vol. 7341, pp. 311–328. Springer, Heidelberg (2012)
34. Rønjom, S., Helleseth, T.: A new attack on the filter generator. IEEE Trans. Inf. Theory **53**(5), 1752–1758 (2007)
35. Rothaus, O.S.: On "bent" functions. J. Comb. Theory **20A**, 300–305 (1976)
36. Tang, D., Carlet, C., Tang, X.: Highly nonlinear boolean functions with optimal algebraic immunity and good behavior against fast algebraic attacks. IEEE Tran. Inf. Theory **59**(1), 653–664 (2013)
37. Tokareva, N.: On the number of bent functions from iterative constructions: lower bounds and hypotheses. Adv. Math. Commun. (AMC) **5**, 609–621 (2011)
38. Wu, B.: $\mathcal{PS}$ bent functions constructed from finite pre-quasifield spreads. http://arxiv.org/abs/1308.3355

Boolean and Vectorial Functions

Some Results on Difference Balanced Functions

Alexander Pott[1] and Qi Wang[2]([✉])

[1] Faculty of Mathematics, Institute of Algebra and Geometry, Otto-von-Guericke University Magdeburg, 39106 Magdeburg, Germany
alexander.pott@ovgu.de
[2] Department of Electrical and Electronic Engineering, South University of Science and Technology of China, Nanshan Shenzhen 518055, Guangdong, China
ust.qiwang@gmail.com,wangqi@sustc.edu.cn

Abstract. For a difference balanced function f from $\mathbb{F}_{q^n}^*$ to $\mathbb{F}_q$, the set $D := \{(x, f(x)) : x \in \mathbb{F}_{q^n}^*\}$ is a generalized difference set with respect to two exceptional subgroups $(\mathbb{F}_{q^n}^*, \cdot)$ and $(\mathbb{F}_q, +)$. This allows us to prove the balance property of difference balanced functions from $\mathbb{F}_{q^n}^*$ to $\mathbb{F}_q$ where q is a prime power. We further prove two necessary and sufficient conditions for d-homogeneous difference balanced functions. This unifies several combinatorial objects related to difference balanced functions.

Keywords: Difference balanced function · d-homogeneous function · Generalized difference set · p-ary sequence with ideal autocorrelation · Two-tuple balanced function

1 Introduction

Throughout this paper, let $q = p^m$ with p an odd prime. A function $f : \mathbb{F}_{q^n}^* \to \mathbb{F}_q$ is said to be *balanced*, if

$$\left|\{x \in \mathbb{F}_{q^n}^* : f(x) = 0\}\right| = q^{n-1} - 1,$$

and

$$\left|\{x \in \mathbb{F}_{q^n}^* : f(x) = b\}\right| = q^{n-1},$$

for each $b \in \mathbb{F}_q^*$. A function $f : \mathbb{F}_{q^n}^* \to \mathbb{F}_q$ is called *difference balanced*, if $f(ax) - f(x)$ is balanced for each $a \in \mathbb{F}_{q^n}^* \setminus \{1\}$. Difference balanced functions arise from p-ary sequences with the ideal two-level autocorrelation. We may naturally define a p-ary sequence $\mathbf{s}$ of period $p^n - 1$ using a function $f : \mathbb{F}_{p^n}^* \to \mathbb{F}_p$ by $s_i := f(\theta^i)$, where θ is a primitive element of $\mathbb{F}_{p^n}$. It was proved in [1] that the function $f : \mathbb{F}_{p^n}^* \to \mathbb{F}_p$ is difference balanced if and only if the sequence $\mathbf{s}$ has the ideal two-level autocorrelation as

$$\mathcal{C}_{\mathbf{s}}(\tau) = \begin{cases} p^n - 1 & \text{if } \tau \equiv 0 \pmod{p^n - 1}, \\ -1 & \text{if } \tau \not\equiv 0 \pmod{p^n - 1}, \end{cases}$$

The work was partially done when the second author was working at Institute of Algebra and Geometry, Otto-von-Guericke University Magdeburg, Magdeburg, Germany, supported by the Alexander von Humboldt Foundation.

Ç. Koç et al. (Eds.): WAIFI 2014, LNCS 9061, pp. 111–120, 2015.
DOI: 10.1007/978-3-319-16277-5_6

where

$$C_{\mathbf{s}}(\tau) = \sum_{0 \le i < p^n - 1} \zeta_p^{s_{i+\tau} - s_i},$$

and ζ_p is a p-th root of unity. In the literature, there are only a few constructions of difference balanced functions from $\mathbb{F}_{q^n}^*$ to $\mathbb{F}_q$:

- the trace function $f(x) = \mathrm{Tr}_{q^n/q}(x)$, where $\mathrm{Tr}_{q^n/q}$ denotes the trace function from $\mathbb{F}_{q^n}$ to $\mathbb{F}_q$;
- the Helleseth-Gong function [2,3]:

$$f(x) = \mathrm{Tr}_{q^n/q}\left(\sum_{i=0}^{\ell} u_i x^{(q^{2ki}+1)/2}\right),$$

where $n = (2\ell + 1)k$, u_i's are defined from b_i's as $u_0 = b_0/2 = (p+1)/2$, $u_i = b_{2i}$ for $i = 1, \ldots, \ell$, and b_i's are defined as $b_0 = 1$, $b_{ij} = (-1)^i$ for an integer j with $1 \le j \le 2\ell$ and $\gcd(j, 2\ell + 1) = 1$, $b_i = b_{2\ell+1-i}$ for $i = 1, \ldots, \ell$ (all the indices of the b_i's are taken modulo $2\ell + 1$);
- the Lin function [4–6] in characteristic 3:

$$f(x) = \mathrm{Tr}_{3^n/3}(x + x^e),$$

where $n \ge 3$ is an odd integer and $e = 2 \cdot 3^{(n-1)/2} + 1$;
- the cascaded composition of the functions above using the Gordon-Mills-Welch method [7] or the method by No [1].

All currently known difference balanced functions listed above are associated with the *d-homogeneity*. A function $f : \mathbb{F}_{q^n}^* \to \mathbb{F}_q$ is called *d-homogeneous*, if there exists an integer d with $\gcd(d, q - 1) = 1$ such that $f(ax) = a^d f(x)$ for all $x \in \mathbb{F}_{q^n}^*$ and each $a \in \mathbb{F}_q^*$. Difference balanced functions are closely related to other combinatorial objects, such as cyclic difference sets, relative difference sets with Singer parameters [1,8–11], and two-tuple balanced functions [12] (see also [13]).

In this paper, we view difference balanced functions as certain generalized difference sets with two exceptional subgroups, which are defined in [14]. Using the group ring representation, we show that every difference balanced function from $\mathbb{F}_{q^n}^*$ to $\mathbb{F}_q$ must be balanced or an affine shift of a balanced function. Furthermore, we establish the equivalence relations between d-homogeneous difference balanced functions and other two objects: two-tuple balanced functions and relative difference sets with Singer parameters.

2 Preliminaries

In this section, we introduce generalized difference sets, which are a generalization of difference sets, and relative difference sets. Difference balanced functions from $\mathbb{F}_{q^n}^*$ to $\mathbb{F}_q$ can then be characterized by generalized difference sets with respect to two exceptional subgroups of orders q and $q^n - 1$. We also give the definition of two-tuple balanced functions, which are in fact equivalent to d-homogeneous difference balanced functions (we will explain this relation in Sect. 4).

For a subset D of a group G, we may identify D with the group ring element $\sum_{g \in G} d_g g \in \mathbb{Z}[G]$, which is also denoted by D (by abuse of notation). Addition and multiplication in group rings are defined as:

$$\sum_{g \in G} a_g g + \sum_{g \in G} b_g g = \sum_{g \in G} (a_g + b_g)g,$$

and

$$\sum_{g \in G} a_g g \sum_{g \in G} b_g g = \sum_{g \in G} \left(\sum_{h \in G} a_h b_{h^{-1}g} \right) g.$$

It is well known that this group ring notation is appropriate to investigate difference properties of certain functions. We will see another application in this paper.

2.1 Generalized Difference Sets

In [14], difference sets were generalized by involving an arbitrary number of exceptional subgroups. Let G be an abelian group of order v (written multiplicatively). Let $N_1, \ldots, N_r$ be subgroups of G of orders $n_1, \ldots, n_r$ and intersecting pairwise trivially. A k-subset D of G is a $(v; n_1, \ldots, n_r; k, \lambda; \lambda_1, \ldots, \lambda_r)$ generalized difference set relative to the subgroups N_i, if the list of differences $d_1 d_2^{-1}$ with $d_1, d_2 \in D$ and $d_1 \neq d_2$, contains each element in $G \setminus (N_1 \cup N_2 \cup \cdots N_r)$ exactly λ times, and every non-identity element in N_i exactly λ_i times. The N_i's are called exceptional subgroups. A generalized difference set D is called cyclic (or abelian) if the group G is cyclic (or abelian).

The (v, k, λ) difference sets are special types of generalized difference sets with parameters $(v; 1; k, \lambda; 0)$, the (m, n, k, λ) relative difference sets are those with parameters $(mn; n; k, \lambda; 0)$, and the $(m, n, k, \lambda_1, \lambda_2)$ divisible difference sets are those with parameters $(mn; n; k, \lambda_2; \lambda_1)$. By convention, we still use (v, k, λ), (m, n, k, λ) and $(m, n, k, \lambda_1, \lambda_2)$ to denote the parameters of difference sets, relative difference sets, and divisible difference sets, respectively. For more information on these combinatorial objects, we refer to [15–18].

Let $H \leq G$ be a subgroup of G, and let ρ_H denote the canonical epimorphism $G \to G/H$ with $\rho_H(g) = gH$. For a generalized difference set D, if $H \leq N_i$ for some N_i with $\lambda_i = 0$, then $\rho_H(D)$ can be interpreted as a subset of G/H. Here we call $\rho_H(D)$ a projection of D. In particular, we have the following result on projections of generalized difference sets with respect to two exceptional subgroups.

Proposition 1. *[17] Let D be a generalized $(v; n_1, n_2; k, \lambda; 0, 0)$ difference set in G relative to N_1 and N_2. If $H \leq N_1$ is a subgroup of order m, then $\rho_H(D)$ is a generalized $(v/m; n_1/m, n_2; k, m\lambda; 0, \lambda(m-1))$ difference set.*

Generalized difference sets with parameters $(n(n-1); n, n-1; n-1, 1; 0, 0)$ in a group G relative to subgroups N_1 of order n and N_2 of order $n-1$ are called *direct product difference sets* [17, 19] (for their connections to projective planes, see also [14]). For a difference balanced function $f : \mathbb{F}_{q^n}^* \to \mathbb{F}_q$, it is easy to prove the following connection between f and the corresponding set $\{(x, f(x)) : x \in \mathbb{F}_{q^n}^*\}$.

Proposition 2. *A function $f : \mathbb{F}_{q^n}^* \to \mathbb{F}_q$ is difference balanced, if and only if the set*

$$D := \{(x, f(x)) : x \in \mathbb{F}_{q^n}^*\} \subseteq G = N_2 \times N_1$$

is a generalized $(q(q^n - 1); q, q^n - 1; q^n - 1, q^{n-1}; 0, q^{n-1} - 1)$ difference set relative to $N_1 = (\mathbb{F}_q, +)$ and $N_2 = (\mathbb{F}_{q^n}^*, \cdot)$. In the language of group rings, in $\mathbb{Z}[G]$ we have,

$$DD^{(-1)} = q^n + q^{n-1}G - q^{n-1}N_1 - N_2, \tag{2.1}$$

where $D^{(-1)}$ denotes the group ring element identifying the subset $\{(x^{-1}, -y) : (x, y) \in D\}$.

By Proposition 1, the trace function from $\mathbb{F}_{q^n}^*$ to $\mathbb{F}_q$ is the first example of difference balanced functions that corresponds to a generalized difference set $\{(x, \mathrm{Tr}_{q^n/q}(x)) : x \in \mathbb{F}_{q^n}^*\}$, which is the projection of the classical direct product difference set $\{(x, x) : x \in \mathbb{F}_{q^n}^*\}$. We will later use the group ring Eq. (2.1) to derive the balance property of difference balanced functions in Sect. 3. Afterwards in Sect. 4, we deal with difference balanced functions with an additional property, i.e., the d-homogeneous property.

2.2   Two-Tuple Balanced Functions

The two-tuple balance property is one of several randomness measurements for sequences, and it is also important for determining the autocorrelation of GMW sequences [12]. We now describe the two-tuple balance property of functions from $\mathbb{F}_{q^n}^*$ to $\mathbb{F}_q$.

Definition 1. *[12] For a function $f : \mathbb{F}_{q^n}^* \to \mathbb{F}_q$, define*

$$N_{b_1, b_2}(a) = |\{x \in \mathbb{F}_{q^n}^* : (f(x), f(ax)) = (b_1, b_2)\}|,$$

where $a \neq 0$ and $b_1, b_2 \in \mathbb{F}_q$. Then f is called two-tuple balanced *if the following conditions are satisfied:*

(i) If $a \notin \mathbb{F}_q^*$, then

$$N_{b_1, b_2}(a) = \begin{cases} q^{n-2} - 1 & \text{if } (b_1, b_2) = (0, 0), \\ q^{n-2} & \text{if } (b_1, b_2) \neq (0, 0). \end{cases}$$

(ii) If $a \in \mathbb{F}_q^*$, then there exists some $\mu \in \mathbb{F}_q$ such that $(f(x), f(ax)) = (b_1, \mu b_1)$ for all $x \in \mathbb{F}_{q^n}^*$ and

$$N_{b_1,b_2}(a) = \begin{cases} q^{n-1} - 1 & \text{if } (b_1, b_2) = (0,0), \\ q^{n-1} & \text{if } (b_1, b_2) = (b_1, \mu b_1) \text{ for } b_1 \in \mathbb{F}_q, \\ 0 & \text{otherwise.} \end{cases}$$

In [13], the following relation between the two-tuple balance property and the difference balance property was given.

Theorem 1. *[13] If a difference balanced function $f : \mathbb{F}_{q^n}^* \to \mathbb{F}_q$ is d-homogeneous for some d with $\gcd(d, q-1) = 1$, then f must be two-tuple balanced.*

Conversely, for a function from $\mathbb{F}_{q^n}^*$ to $\mathbb{F}_q$, it is easy to see that the two-tuple balance property implies both the difference balance property and the balance property. In Sect. 4, we will see that the converse of Theorem 1 also holds, i.e., the two-tuple balance property implies the d-homogeneity.

3   The Balance Property of Difference Balanced Functions

It was proved in [20] that every difference balanced function f from $\mathbb{F}_{q^n}^*$ to $\mathbb{F}_q$ must be balanced or an affine shift of a balanced function g, if q is a prime. Now we show that this is also true for q a prime power. We note that the proof techniques in [20] for the q prime case cannot be adapted here, because the additive character group of $\mathbb{F}_q$ for q prime power is not cyclic. In comparison, the proof of Theorem 2 is more transparent by counting the appearances of certain elements in the group ring Eq. (2.1).

Theorem 2. *Every difference balanced function $f : \mathbb{F}_{q^n}^* \to \mathbb{F}_q$ must be of the form $f = g + b$ for some $b \in \mathbb{F}_q$, where the function g is balanced.*

Proof. Define $D_b := \{(x, b) \in G : f(x) = b\}$ and $d_b = |D_b|$, where $G = \mathbb{F}_{q^n}^* \times \mathbb{F}_q$ and $b \in \mathbb{F}_q$. We count the coefficient of certain elements in $DD^{(-1)}$ in two ways: one by direct calculation, and the other by the group ring Eq. (2.1).

First, by counting the coefficient of $(a, 0)$ in $DD^{(-1)}$ where $a \in \mathbb{F}_{q^n}^*$, we have

$$\sum_{b \in \mathbb{F}_q} d_b^2 = q^n + q^{n-1}(q^n - 1) - q^{n-1} - (q^n - 1)$$

$$= q^{2n-1} - 2q^{n-1} + 1. \tag{3.1}$$

Second, we calculate the coefficient of (a, b) in $DD^{(-1)}$ where $a \in \mathbb{F}_{q^n}^*$ and $b \neq 0$:

$$\sum_{\substack{(b_1, b_2) \\ b_1 \neq b_2}} 2 d_{b_1} d_{b_2} = q^{n-1}(q^n - 1)(q - 1) - q^{n-1}(q - 1)$$

$$= q^{2n} - 2q^n - q^{2n-1} + 2q^{n-1}. \tag{3.2}$$

With (3.1) and (3.2), we get

$$(q-1)\sum_{\substack{b\in\mathbb{F}_q}} d_b^2 - \sum_{\substack{(b_1,b_2)\\b_1\neq b_2}} 2d_{b_1}d_{b_2} = \sum_{\substack{(b_1,b_2)\\b_1\neq b_2}} (d_{b_1}-d_{b_2})^2 = q-1. \tag{3.3}$$

Since the summation of $(d_{b_1}-d_{b_2})^2$ is not 0, there exist at least two distinct elements $b_1, b_2 \in \mathbb{F}_q$ such that $|d_{b_1}-d_{b_2}| \geq 1$. Without loss of generality, suppose that there are k elements $b \in \mathbb{F}_q$ with $d_b = d_{b_1}$, where $1 \leq k \leq q-1$. Then there are $q-k$ elements $b \in \mathbb{F}_q$ left with $d_b \neq d_{b_1}$ (these $q-k$ of d_b may or may not equal to d_{b_2}). Thus, we have

$$q-1 = \sum_{\substack{(b_1,b_2)\\b_1\neq b_2}} (d_{b_1}-d_{b_2})^2 \geq k(q-k),$$

which shows that $k \geq q-1$ or $k \leq 1$. It then follows that (3.3) is satisfied if and only if $d_c = \hat{d}$ for some $c \in \mathbb{F}_q$ and $d_b = \hat{d}+1$ or $d_b = \hat{d}-1$ for each $b \neq c$. Note that $\sum_{b\in\mathbb{F}_q} d_b = q^n - 1$, hence we have $d_c = q^{n-1} - 1$ for some $c \in \mathbb{F}_q$ and $d_b = q^{n-1}$ for each $b \neq c$. The proof is then completed.

Remark 1. By Theorem 2, we see that every difference balanced function must be balanced, or an affine shift of a balanced function. Without loss of generality, we may always assume that a difference balanced function f is balanced (otherwise, replace f by $f - b$ for a suitable $b \in \mathbb{F}_q^*$).

4   *d*-Homogeneous Difference Balanced Functions

In this section, we give two necessary and sufficient conditions for d-homogeneous difference balanced functions from $\mathbb{F}_{q^n}^*$ to $\mathbb{F}_q$.

Theorem 3. *For a function f from $\mathbb{F}_{q^n}^*$ onto $\mathbb{F}_q$, the following three statements are equivalent:*

(i) *f is difference balanced and d-homogeneous for some d with $\gcd(d, q-1) = 1$;*

(ii) *f is two-tuple balanced;*

(iii) *for each $b \in \mathbb{F}_q^*$, the set $C := \{x \in \mathbb{F}_{q^n}^* : f(x) = b\}$ is a relative difference set with parameters $\left(\frac{q^n-1}{q-1}, q-1, q^{n-1}, q^{n-2}\right)$.*

Proof.   – (i) $\Rightarrow$ (ii): See Theorem 1.

– (ii) $\Rightarrow$ (i): Since the two-tuple balance property implies the difference balance property, it suffices to prove that f is d-homogeneous. By the definition of the two-tuple balance property, for each $a \in \mathbb{F}_q^*$, there exists some $\mu_a \in \mathbb{F}_q^*$ such that $f(ax) = \mu_a f(x)$ for all $x \in \mathbb{F}_{q^n}^*$. It is clear that $\mu_a = 1$ if $a = 1$. On the other hand, since f is difference balanced, by Proposition 2, D is a generalized difference set satisfying (2.1). If $\mu_a = 1$, suppose that $a \neq 1$. We consider

$$D = \{(x, f(x)) : x \in \mathbb{F}_{q^n}^*\} = \{(x, f(ax)) : x \in \mathbb{F}_{q^n}^*\}.$$

Then there are $q^{n-1} - 1$ appearances of $(1,0)$ in $(1, f(ax) - f(x))$ of $DD^{(-1)}$ with $a \neq 1$, but this is a contradiction to (2.1). Thus, it follows that $\mu_a = 1$ if and only if $a = 1$. Similarly, we further obtain that $a = a'$ if and only if $\mu_a = \mu_{a'}$. Define a mapping $\epsilon : a \mapsto \mu_a$, we then have $\epsilon(aa') = \mu_a\mu_{a'} = \epsilon(a)\epsilon(a')$. Thus, ϵ is indeed an automorphism of $\mathbb{F}_q^*$. This means there must exist a d with $\gcd(d, q-1) = 1$ such that $\mu_a = a^d$, and then f is d-homogeneous.

- (ii) $\Rightarrow$ (iii): Define

$$D_b := \{x \in \mathbb{F}_{q^n}^* : f(x) = b\},$$

where $b \in \mathbb{F}_q^*$. Then by the definition of the two-tuple balance property, we have

$$D_b D_b^{(-1)} = q^{n-1} + q^{n-2}\left(\mathbb{F}_{q^n}^* - \mathbb{F}_q^*\right)$$
$$= q^{n-1} + q^{n-2}\mathbb{F}_{q^n}^* - q^{n-2}\mathbb{F}_q^*,$$

which means that D_b is a difference set in $\mathbb{F}_{q^n}^*$ relative to $\mathbb{F}_q^*$ with parameters $\left(\frac{q^n-1}{q-1}, q-1, q^{n-1}, q^{n-2}\right)$.

- (iii) $\Rightarrow$ (ii): Suppose that $C \subseteq \mathbb{F}_{q^n}^*$ is a relative difference set with the parameters described in (iii). Define C_b as

$$C_b := \{bx : x \in C\},$$

for all $b \in \mathbb{F}_q^*$, and $C_0 := \mathbb{F}_{q^n}^* \setminus \mathbb{F}_q^* C$. Then $\mathbb{F}_{q^n}^*$ is the disjoint union of all the C_b's for $b \in \mathbb{F}_q$. We then define a function $f : \mathbb{F}_{q^n}^* \to \mathbb{F}_q$ as

$$f(x) := b^d, \text{ if } x \in C_b, \text{where } \gcd(d, q-1) = 1.$$

We now show that f is two-tuple balanced. Since C is a $((q^n - 1)/(q-1), q-1, q^{n-1}, q^{n-2})$ difference set relative to $\mathbb{F}_q^*$, we have the group ring equation:

$$CC^{(-1)} = q^{n-1} + q^{n-2}\mathbb{F}_{q^n}^* - q^{n-2}\mathbb{F}_q^*.$$

It then follows that

$$\begin{aligned}
C_0 C_0^{(-1)} &= (\mathbb{F}_{q^n}^* - \mathbb{F}_q^* C)(\mathbb{F}_{q^n}^* - \mathbb{F}_q^* C^{(-1)}) \\
&= \mathbb{F}_{q^n}^* \mathbb{F}_{q^n}^* - \mathbb{F}_q^* \mathbb{F}_{q^n}^* C^{(-1)} - \mathbb{F}_q^* \mathbb{F}_{q^n}^* C + \mathbb{F}_q^* \mathbb{F}_q^* CC^{(-1)} \\
&= (q^n - 1)\mathbb{F}_{q^n}^* - 2(q-1)q^{n-1}\mathbb{F}_{q^n}^* + (q-1)\mathbb{F}_q^* CC^{(-1)} \\
&= (-q^n + 2q^{n-1} - 1)\mathbb{F}_{q^n}^* \\
&\quad + (q-1)\mathbb{F}_q^*(q^{n-1} + q^{n-2}\mathbb{F}_{q^n}^* - q^{n-2}\mathbb{F}_q^*) \\
&= (q^{n-1} - 1) + (q^{n-1} - 1)(\mathbb{F}_q^* - \{1\}) \\
&\quad + (q^{n-2} - 1)(\mathbb{F}_{q^n}^* - \mathbb{F}_q^*). \quad (4.1)
\end{aligned}$$

Similarly, for $b \in \mathbb{F}_q^*$, we have

$$C_b C_b^{(-1)} = q^{n-1} + q^{n-2}(\mathbb{F}_{q^n}^* - \mathbb{F}_q^*); \quad (4.2)$$

for $b \in \mathbb{F}_q^*$, we have

$$C_0 C_b^{(-1)} = q^{n-2}(\mathbb{F}_{q^n}^* - \mathbb{F}_q^*) \tag{4.3}$$

and

$$C_b C_0^{(-1)} = q^{n-2}(\mathbb{F}_{q^n}^* - \mathbb{F}_q^*). \tag{4.4}$$

For $b_1, b_2 \in \mathbb{F}_q^*$ with $b_1 \neq b_2$, we have

$$C_{b_1} C_{b_2}^{(-1)}$$
$$= q^{n-1}(b_1 b_2^{-1}) + q^{n-2}(\mathbb{F}_{q^n}^* - \mathbb{F}_q^*). \tag{4.5}$$

In fact, (4.1)–(4.5) are equivalent to the two-tuple balance property of the function f. For instance, by (4.5), we have

$$N_{b_1,b_2}(a) = \begin{cases} q^{n-1} & \text{if } a^d = b_1 b_2^{-1}, \\ q^{n-2} & \text{if } a \notin \mathbb{F}_q^*, \\ 0 & \text{otherwise.} \end{cases}$$

Similarly, one can verify that f is two-tuple balanced by (4.1)–(4.5). The proof is then completed.

Remark 2. By (i) and (iii) in Theorem 3, if $f : \mathbb{F}_{q^n}^* \to \mathbb{F}_q$ is d-homogeneous and difference balanced, we see that the preimage set of the function f

$$D_b := \{x \in \mathbb{F}_{q^n}^* : f(x) = b\}$$

is a $\left(\frac{q^n-1}{q-1}, q-1, q^{n-1}, q^{n-2}\right)$ difference set relative to $\mathbb{F}_q^*$ for each $b \in \mathbb{F}_q^*$, and

$$D_0 := \{x \in \mathbb{F}_{q^n}^* : f(x) = 0\}$$

is actually a divisible difference set with parameters

$$\left(\frac{q^n - 1}{q - 1}, q - 1, q^{n-1} - 1, q^{n-1} - 1, q^{n-2} - 1\right).$$

By Proposition 1, via the canonical epimorphism $\rho : \mathbb{F}_{q^n}^* \to \mathbb{F}_{q^n}^*/\mathbb{F}_q^*$, for each $b \in \mathbb{F}_q^*$, the projection $\rho(D_b)$ is a cyclic difference set with Singer parameters

$$\left(\frac{q^n - 1}{q - 1}, q^{n-1}, q^{n-2}(q - 1)\right),$$

and $\rho(D_0)$ is the complement of $\rho(D_b)$, i.e., a cyclic difference set with Singer parameters

$$\left(\frac{q^n - 1}{q - 1}, \frac{q^{n-1} - 1}{q - 1}, \frac{q^{n-2} - 1}{q - 1}\right).$$

This unifies some of the results in [1,8,11].

Since the connection between (i) and (iii) in Theorem 3 has not been observed before, the original proofs in [1,17] had been involved and their similarity, according to the following corollary, had not been unveiled. The Gordon-Mills-Welch construction [7] and the No construction [1] can be reformulated as the product of two relative difference sets as follows.

Corollary 1. *[17, Corollary 3.2.2] Suppose that D_1 is a difference set in $\mathbb{F}_{q^n}^*$ relative to $\mathbb{F}_{q^\ell}^*$ with parameters $\left(\frac{q^n-1}{q^\ell-1}, q^\ell - 1, q^{n-\ell}, q^{n-2\ell}\right)$, where ℓ is a divisor of n, and D_2 is a $\left(\frac{q^\ell-1}{q-1}, q - 1, q^{\ell-1}, q^{\ell-2}\right)$ difference set in $\mathbb{F}_{q^\ell}^*$ relative to $\mathbb{F}_q^*$. Then the set*

$$D := \{d_1 d_2 : d_1 \in D_1, d_2 \in D_2\}$$

is a difference set in $\mathbb{F}_{q^n}^*$ relative to $\mathbb{F}_q^*$ with parameters

$$\left(\frac{q^n - 1}{q - 1}, q - 1, q^{n-1}, q^{n-2}\right).$$

References

1. No, J.S.: New cyclic difference sets with Singer parameters constructed from d-homogeneous functions. Des. Codes Cryptogr. **33**(3), 199–213 (2004)
2. Helleseth, T., Kumar, P.V., Martinsen, H.: A new family of ternary sequences with ideal two-level autocorrelation function. Des. Codes Cryptogr. **23**(2), 157–166 (2001)
3. Helleseth, T., Gong, G.: New nonbinary sequences with ideal two-level autocorrelation. IEEE Trans. Inform. Theory **48**(11), 2868–2872 (2002)
4. Lin, H.A.: From cyclic Hadamard difference sets to perfectly balanced sequences. Ph.D. thesis, University of Southern California (1998)
5. Arasu, K., Dillon, J., Player, K.: Character sum factorizations yield perfect sequences. Preprint (2010)
6. Hu, H., Shao, S., Gong, G., Helleseth, T.: The proof of Lin's conjecture via the decimation-hadamard transform (2013). arXiv preprint arXiv:1307.0885
7. Gordon, B., Mills, W.H., Welch, L.R.: Some new difference sets. Canad. J. Math. **14**, 614–625 (1962)
8. Chandler, D.B., Xiang, Q.: Cyclic relative difference sets and their p-ranks. Des. Codes Cryptogr. **30**(3), 325–343 (2003)
9. Chandler, D.B., Xiang, Q.: The invariant factors of some cyclic difference sets. J. Combin. Theory Ser. A **101**(1), 131–146 (2003)
10. No, J.S., Shin, D.J., Helleseth, T.: On the p-ranks and characteristic polynomials of cyclic difference sets. Des. Codes Cryptogr. **33**(1), 23–37 (2004)
11. Kim, S.H., No, J.S., Chung, H., Helleseth, T.: New cyclic relative difference sets constructed from d-homogeneous functions with difference-balanced property. IEEE Trans. Inform. Theory **51**(3), 1155–1163 (2005)
12. Golomb, S.W., Gong, G.: Signal Design for Good Correlation: For Wireless Communication, Cryptography, and Radar. Cambridge University Press, Cambridge (2005)
13. Gong, G., Song, H.Y.: Two-tuple balance of non-binary sequences with ideal two-level autocorrelation. Discrete Appl. Math. **154**(18), 2590–2598 (2006)

14. Pott, A., Wang, Q., Zhou, Y.: Sequences and functions derived from projective planes and their difference sets. In: Özbudak, F., Rodríguez-Henríquez, F. (eds.) WAIFI 2012. LNCS, vol. 7369, pp. 64–80. Springer, Heidelberg (2012)
15. Beth, T., Jungnickel, D., Lenz, H.: Design Theory. Encyclopedia of Mathematics and its Applications, vol. 78, 2nd edn. Cambridge University Press, Cambridge (1999)
16. Lander, E.S.: Symmetric Designs: An Algebraic Approach. London Mathematical Society Lecture Note Series, vol. 74. Cambridge University Press, Cambridge (1983)
17. Pott, A.: Finite Geometry and Character Theory. Lecture Notes in Mathematics, vol. 1601. Springer, Heidelberg (1995)
18. Schmidt, B.: Characters and Cyclotomic Fields in Finite Geometry. Lecture Notes in Mathematics, vol. 1797. Springer, Heidelberg (2002)
19. Ganley, M.J.: Direct product difference sets. J. Comb. Theory Ser. A **23**(3), 321–332 (1977)
20. Ludkovski, M., Gong, G.: Ternary ideal 2-level autocorrelation sequences. CORR 2000–59, Techinical report of CACR, University of Waterloo (2000)

Affine Equivalency and Nonlinearity Preserving Bijective Mappings over $\mathbb{F}_2$

İsa Sertkaya[1,2]([⊠]), Ali Doğanaksoy[1,3], Osmanbey Uzunkol[2]([⊠]),
and Mehmet Sabır Kiraz[2]

[1] Institute of Applied Mathematics, Middle East Technical University,
Ankara, Turkey
aldoks@metu.edu.tr
[2] Mathematical and Computational Sciences, TÜBİTAK BİLGEM UEKAE,
Gebze, Kocaeli, Turkey
{isa.sertkaya,osmanbey.uzunkol,mehmet.kiraz}@tubitak.gov.tr
[3] Department of Mathematics, Middle East Technical University, Ankara, Turkey

Abstract. We first give a proof of an isomorphism between the group of affine equivalent maps and the automorphism group of Sylvester Hadamard matrices. Secondly, we prove the existence of new nonlinearity preserving bijective mappings without explicit construction. Continuing the study of the group of nonlinearity preserving bijective mappings acting on n-variable Boolean functions, we further give the exact number of those mappings for $n \leq 6$. Moreover, we observe that it is more beneficial to study the automorphism group of bijective mappings as a subgroup of the symmetric group of the 2^n dimensional $\mathbb{F}_2$-vector space due to the existence of non-affine mapping classes.

Keywords: Cryptographic Boolean functions · Affine equivalence · Nonlinearity preserving mappings · Sylvester Hadamard matrices

1 Introduction

A natural way to study and analyze a very large algebraic set is to first partition it under an equivalence relation, next choose a representative for each class and then analyze the reduced sized set composing of these representatives. Such a procedure is vital for the set of Boolean functions due to their applications varying from switching theory, coding theory to cryptography.

The study of the actions of transformations on Boolean functions dates back to Harrison [13,14], and later [8,29,30] where the main concern was their application in switching theory. In coding theory, affine transformations are analyzed especially for the Reed-Muller codes [15,18,19].

Studying equivalence of Boolean functions, more precisely nonlinearity preserving mappings, has both theoretical and practical influences in cryptographic

---

A preliminary version can be found in [28] at Cryptology ePrint Archieve.

Ç. Koç et al. (Eds.): WAIFI 2014, LNCS 9061, pp. 121–136, 2015.
DOI: 10.1007/978-3-319-16277-5_7

research. Searching Boolean functions possessing higher cryptographic design criteria values is an ongoing research subject. The aim is to find families of Boolean functions not equivalent to the existing ones which lead to obtain better primitives for cryptographic design. Indeed without loosing the security margins, choosing efficiently implementable Boolean function by using these mappings is an ultimate research goal.

For symmetric cryptosystems, Meier and Staffelbach in [20] showed that nonlinearity is invariant under the action of general affine group of degree n over $\mathbb{F}_2$, AGL_n. By extending the result of [18, p. 417], it is proved that affine equivalence relations ($\mathsf{AGL}_n \ltimes \mathcal{A}_n$) also preserve nonlinearity [22]. In 1998, Carlet, Charpin and Zinoviev proposed CCZ-equivalence for S-boxes, but later Budaghyan and Carlet proved that for Boolean functions, CCZ-equivalence reduces to affine equivalence, see [3,4] respectively. Note that for cryptographic applications counting the Boolean functions having the same nonlinearity value may become feasible even for larger n values, see [16].

By considering the truth tables, the set of all Boolean functions can be seen as the $\mathbb{F}_2-$vector space $\mathbb{F}_2^{2^n}$. Hence, the set of extended transformations acting on n variable Boolean functions to all bijective transformations defined over $\mathbb{F}_2^{2^n}$ (i.e., to the symmetric group $\mathcal{S}_{2^{2^n}}$) is a reasonable object to study. Therefore, the problem becomes to determine the subgroup of $\mathcal{S}_{2^{2^n}}$ that preserves the nonlinearity for all Boolean functions. In this context, the so-called affine equivalence relations $\mathsf{AGL}_n \ltimes \mathcal{A}_n$ correspond indeed only to a subgroup of general affine group of degree 2^n over $\mathbb{F}_2$, namely AGL_{2^n}. Even if the affine equivalence relations $\mathsf{AGL}_n \ltimes \mathcal{A}_n$ are widely studied [1,2,10], this interesting yet very challenging problem still remains open, and surprisingly it does not get much attention in the literature.

In the pursue of determining the nonlinearity preserving transformations, the authors in [25,26] first mapped the action of $\mathsf{AGL}_n \ltimes \mathcal{A}_n$ from the algebraic normal forms to an action on the truth tables, and then proved those well-known affine equivalence relation properties analogously using the truth tables and signed sequences of the functions. Later it is shown that there exists a correspondence between AGL_n and the automorphism group of Sylvester Hadamard matrices $\mathsf{Aut}(H_n)$ [27]. Furthermore, the existence of nonlinearity preserving non-affine bijective transformations is shown [27].

1.1 Our Contributions

We first revisit the group of all affine equivalence relations. We give a proof that the correspondence given in [27] is actually an isomorphism between the automorphism group of Sylvester Hadamard matrices and the group of affine equivalence relations, i.e. $\mathsf{Aut}(H_n) \cong \mathsf{AGL}_n \ltimes \mathcal{A}_n$. Hence, we determine the cardinality of $\mathsf{Aut}(H_n)$ by using counting techniques for AGL_n and $\mathcal{A}_n$. Furthermore, without constructing explicitly, we prove the existence of a new class of nonlinearity preserving transformations based on the cardinalities of nonlinearity class partitions. Following these results, we analyze the notion of automorphism group of nonlinearity classes and propose that studying the automorphism group of nonlinearity classes as a subgroup of $\mathcal{S}_{2^{2n}}$ is more insightful for cryptographic design purposes, since it contains transformations which do not belong to the group of affine equivalence relations.

Roadmap. In Sect. 2, we give notations and preliminaries. In Sect. 3, we prove that the group of affine equivalence relations is isomorphic to the automorphism group of Sylvester Hadamard matrices. In Sect. 4, we give the exact number of nonlinearity preserving bijective mappings for $n \leq 6$, prove existence of new nonlinearity preserving non-affine mappings, and give some new examples. We discuss the main concerns about the automorphism group of nonlinearity classes in Sect. 5. Section 6 concludes the paper and points out open problems.

2   Preliminaries

Unless otherwise stated the definitions and results of this section can be found at [6, Chap. 8].

Let $\mathbb{F}_2^n$ be the n-dimensional $\mathbb{F}_2$−vector space admitting the usual *lexico-graphical* ordering,

$$\alpha_0 = (0, 0, \ldots, 0) < \alpha_1 = (0, 0, \ldots, 0, 1) < \ldots < \alpha_{2^n-1} = (1, 1, \ldots, 1).$$

Any function $f : \mathbb{F}_2^n \to \mathbb{F}_2$ is called a *Boolean function* and can be represented uniquely by

- the corresponding *truth table* given by

$$T_f = (f(\alpha_0), f(\alpha_1), \ldots, f(\alpha_{2^n-1})),$$

where T_f is lexicographically ordered or,
- the corresponding *algebraic normal form* that is the multivariate polynomial over $\mathbb{F}_2$ given by

$$f(x_n, x_{n-1}, \ldots, x_1) = c_0 \oplus c_1 x_1 \oplus \cdots \oplus c_n x_n \oplus c_{12} x_1 x_2 \oplus \cdots \oplus c_{12\ldots n} x_1 x_2 \cdots x_n.$$

The set of all Boolean functions defined on $\mathbb{F}_2^n$ is denoted by $\mathcal{F}_n$. The *degree* $deg(f)$ of f is the degree of its algebraic normal form, i.e. the number of variables in the highest order term with nonzero coefficient. If $deg(f) \leq 1$ then f is called *affine*, i.e., it is of the form $f(x_n, x_{n-1}, \ldots, x_1) = \langle c, x \rangle \oplus c_0$, where $c_0 \in \mathbb{F}_2$ and $\langle c, x \rangle = c_1 x_1 \oplus \cdots \oplus c_n x_n$ is the *standard inner product* with $x, c \in \mathbb{F}_2^n$. The set of all affine Boolean functions on $\mathbb{F}_2^n$ is denoted by $\mathcal{A}_n$, where each function in $\mathcal{A}_n$ will be given by

$$\ell_{(\alpha_i, a)} : x \mapsto \langle x, \alpha_i \rangle \oplus a.$$

For a given function f, the *Hamming weight* $w(f)$ is given as usual as the number of ones in T_f. The set $Supp(f) := \{\alpha \in \mathbb{F}_2^n \mid f(\alpha) = 1\}$ is called the *support* of the function f. For given functions f and g, the *Hamming distance* $d(f, g)$ is the Hamming weight of $f \oplus g$.

An $n \times n$ matrix H over $\mathbb{Q}$ with all entries ± 1 is called a *Hadamard matrix* if $H \cdot H^t = n I_n$ holds, where I_n is the identity matrix of order n [11]. Hadamard matrices were actually first given by Sylvester [31], which are $2^n \times 2^n$ *Sylvester Hadamard matrices* constructed iteratively by the *Kronecker product* $\otimes$ as follows:

$$H_n = \begin{bmatrix} 1 & 1 \\ 1 & -1 \end{bmatrix} \otimes H_{n-1} = \begin{bmatrix} H_{n-1} & H_{n-1} \\ H_{n-1} & -H_{n-1} \end{bmatrix}.$$

The *Walsh transform* $W_f : \mathbb{F}_2^n \to \mathbb{R}$, of a function f is given by

$$W_f(\omega) = \sum_{x \in \mathbb{F}_2^n} (-1)^{f(x) \oplus \langle x, \omega \rangle} \text{ for } \omega \in \mathbb{F}_2^n.$$

This transformation enables us to map the image of the functions under the group isomorphy $\mathbb{F}_2^n \to (\mathbb{Z}^\star)^n$ to the spectral domain. The truth table of the Walsh transform

$$W_f = (W_f(\alpha_0), W_f(\alpha_1) \ldots, W_f(\alpha_{2^n-1}))$$

is called the *Walsh Spectrum* of f, and it can be expressed as $W_f = \zeta_f H_n$ (see [24]) where

$$\zeta_f = \left((-1)^{f(\alpha_0)}, (-1)^{f(\alpha_1)}, \ldots, (-1)^{f(\alpha_{2^n-1})} \right)$$

is the truth table of the *signed function* $(-1)^{f(x)}$ of f.

Definition & Proposition 1. *Let $\mathcal{S}_n$ be the group of all permutation matrices of order n and $\mathcal{D}_n$ be the group of all diagonal matrices of order n with having only non-zero values at diagonal entries equal to 1 or -1. Then, the elements of the semi-direct product $\mathcal{S}_n^\pm := \mathcal{S}_n \ltimes \mathcal{D}_n$ are called monomial matrices. $\mathcal{S}_n^\pm$ forms a group under the operation $\cdot$ given by*

$$P_1 \cdot P_2 = P_1' P_2' D_1 P_1' D_2 (P_1')^{-1}$$

where $P_1 = P_1' D_1$ and $P_2 = P_2' D_2$ with $P_1', P_2' \in \mathcal{S}_n$ and $D_1, D_2 \in \mathcal{D}_n$.

Proof. It follows immediately by rephrasing the result in [17, Chap. 1, p. 260]. □

Two $n \times n$ Hadamard matrices H' and H'' are called equivalent if H' can be obtained from H'' by performing a finite sequence of permuting the rows, or the columns, or multiplying rows or columns by -1 operations. This induces a group action[1] of the group $\mathcal{S}_n^\pm \times \mathcal{S}_n^\pm$ on the set of $n \times n$ Hadamard matrices [12]. In particular, for a given Hadamard matrix H, the stabilizer of H under the above action of $\mathcal{S}_n^\pm \times \mathcal{S}_n^\pm$ is called an *automorphism* of the Hadamard matrix H. More formally, we have the following definition and proposition.

---

[1] Actually, the action can only be defined for a suitable subgroup of $\mathcal{S}_n^\pm \times \mathcal{S}_n^\pm$ for which the closedness of the map is assured.

Definition & Proposition 2. [12][2] *The direct product $\mathcal{S}_{2^n}^{\pm} \times \mathcal{S}_{2^n}^{\pm}$ acts on the set of Sylvester Hadamard matrices as $(P,Q)H_n := PH_nQ^{-1}$, where $P \in \mathcal{S}_{2^n}^{\pm}$ corresponding to signed row operations and $Q \in \mathcal{S}_{2^n}^{\pm}$ corresponding to signed column operations. Then for each Sylvester Hadamard matrix H_n, the stabilizer of H_n which is the subgroup of $\mathcal{S}_{2^n}^{\pm} \times \mathcal{S}_{2^n}^{\pm}$ that fixes H_n is said to be the automorphism group of H_n and denoted by $\mathsf{Aut}(H_n)$, i.e.,*

$$\mathsf{Aut}(H_n) = \{(P,Q) \in \mathcal{S}_{2^n}^{\pm} \times \mathcal{S}_{2^n}^{\pm}| \ PH_nQ^{-1} = H_n\}.$$

$\mathsf{Aut}(H_n)$ forms a group under the operation $\star$ given by

$$(P_1,Q_1) \star (P_2,Q_2) := (P_1 \cdot P_2, Q_1 \cdot Q_2).$$

Definition & Proposition 3. [21,24] *The nonlinearity N_f of a function f is its distance to the nearest affine function. N_f can be expressed with the Walsh transform of f as follows:*

$$N_f = \min_{\ell_{(\beta,a)} \in \mathcal{A}_n} d(f, \ell_{(\beta,a)}) = 2^{n-1} - \frac{1}{2} \max_{\omega \in \mathbb{F}_2^n} |W_f(\omega)|.$$

A function f is called a *bent function* if $W_f(w) = \pm 2^{n/2}$ for any $w \in \mathbb{F}_2^n$. Bent functions only exist if n is even [9,23], and form a special set $\mathcal{B}_n$ of Boolean functions attaining maximal nonlinearity for a fixed positive even integer n [20].

3   Affine Equivalence

We start with constructing the group which only consists of affine equivalences.

Definition 1. [19] *Let GL_n be the group of all nonsingular matrices of order n over $\mathbb{F}_2$. Let further AGL_n be the group*

$$\mathsf{GL}_n \ltimes \mathbb{F}_2^n := \{(A,\alpha)| \ A \in \mathsf{GL}_n, \ \alpha \in \mathbb{F}_2^n\},$$

with respect to the operation $\bullet$. The group law and its inverse are given by

$$(A,\alpha) \bullet (A',\alpha') := (A'A, \alpha'A + \alpha),$$
$$(A,\alpha)^{-1} := (A^{-1}, \alpha A^{-1})$$

for all $(A,\alpha), (A',\alpha') \in \mathsf{AGL}_n$.
    Similarly, we define the group

$$\mathsf{AGL}_n \ltimes \mathcal{A}_n := \{(\tau, \ell_{(\beta,a)})| \ \tau \in \mathsf{AGL}_n, \ \ell_{(\beta,a)} \in \mathcal{A}_n\},$$

which can be given more explicitly as follows:

$$\mathsf{AGL}_n \ltimes \mathcal{A}_n := \{(A,\alpha,\beta,a) \ | \ A \in \mathsf{GL}_n, \ \alpha,\beta \in \mathbb{F}_2^n, \ a \in \mathbb{F}_2\},$$

where τ and $\ell_{(\beta,a)}$ are given by the mappings $\tau : x \mapsto xA \oplus \alpha$ and $\ell_{(\beta,a)} : x \mapsto \langle x,\beta \rangle \oplus a$.

---

[2] This statement holds for any Hadamard matrices but we only consider the automorphism group of the Sylvester Hadamard matrices.

To the best of our knowledge, although $\mathsf{AGL}_n \ltimes \mathcal{A}_n$ forms a group, its proof is not given explicitly in the literature. For the sake of completeness, we give an elementary proof that $\mathsf{AGL}_n \ltimes \mathcal{A}_n$ forms a group.

Proposition 1. $\mathsf{AGL}_n \ltimes \mathcal{A}_n$ *forms a group under the operation* $\circ$ *and its inverse given by*

$$(\tau, \ell_{(\beta,a)}) \circ (\tau', \ell_{(\beta',a')}) := (\tau \bullet \tau', \tau(\ell_{(\beta',a')}) + \ell_{(\beta,a)}),$$
$$(\tau, \ell_{(\beta,a)})^{-1} := (\tau^{-1}, \tau^{-1}(\ell_{(\beta,a)}))$$

for all $\tau, \tau' \in \mathsf{AGL}_n$ *and* $\ell_{(\beta,a)}, \ell_{(\beta',a')} \in \mathcal{A}_n$, *respectively. This operation and its inverse can also be given by*

$$(A, \alpha, \beta, a) \circ (A', \alpha', \beta', a') := (A'A, \alpha'A \oplus \alpha, \beta'A^t \oplus \beta, \langle \alpha, \beta' \rangle \oplus a' \oplus a),$$
$$(A, \alpha, \beta, a)^{-1} := (A^{-1}, \alpha A^{-1}, \beta(A^{-1})^t, \langle \alpha, \beta(A^{-1})^t \rangle \oplus a)$$

for all $A, A' \in \mathsf{GL}_n$, $\alpha, \alpha', \beta, \beta' \in \mathbb{F}_2^n$, $a, a' \in \mathbb{F}_2$, *respectively.*

Proof. $(A = I_n, \alpha = \alpha_0, \beta = \alpha_0, a = 0)$ is the identity element of $\mathsf{AGL}_n \ltimes \mathcal{A}_n$. Closedness and existence of inverse elements follows immediately from the same properties of the underlying groups of components. Associativity follows also from the associativity of the underlying operations for each component. $\square$

We have an immediate action of the group $\mathsf{AGL}_n \ltimes \mathcal{A}_n$ on $\mathcal{F}_n$ as follows

$$(A, \alpha, \beta, a)f := f(xA \oplus \alpha) \oplus \langle x, \beta \rangle \oplus a.$$

Two Boolean functions f and g are said to be affine equivalent if and only if for $(A, \alpha, \beta, a) \in \mathsf{AGL}_n \ltimes \mathcal{A}_n$ the following holds

$$f(x) = (A, \alpha, \beta, a)g = g(xA \oplus \alpha) \oplus \langle x, \beta \rangle \oplus a.$$

If f and g are affine equivalent functions we denote them by $f \sim g$.

In the following proposition, Preneel proved that the action of an affine equivalence relation results in a signed permutation on the Walsh spectra of the function and consequently preserves its nonlinearity. Furthermore, under the action of $AGL_n \ltimes \mathcal{A}_n$, the distribution of absolute Walsh spectra, nonlinearity and algebraic degree remain invariant [2].

Proposition 2. [22] *Let* $f, g \in \mathcal{F}_n$ *be* $f \sim g$ *with* $f(x) = g(xA \oplus \alpha) \oplus \langle x, \beta \rangle \oplus a$ *then for the Walsh transforms of* f *and* g *the following relation holds:*

$$W_f(\omega) = (-1)^{\langle \alpha, (\omega \oplus \beta)(A^{-1})^t \rangle \oplus a} W_g((\omega \oplus \beta)(A^{-1})^t).$$

Now, we have the following theorem that gives a $1 - 1$ correspondence between $\mathsf{AGL}_n \ltimes \mathcal{A}_n$ and $\mathsf{Aut}(H_n)$.

Theorem 1. *Let the notations be as above and* $f, g \in \mathcal{F}_n$. *Then the following assertions are equivalent:*

(i) There exists a unique $(A, \alpha, \beta, a) \in \mathsf{AGL}_n \ltimes \mathcal{A}_n$ such that $(A, \alpha, \beta, a)g = f$.
(ii) There exists a unique $(P, Q) \in \mathsf{Aut}(H_n)$ such that $\zeta_f = \zeta_g P$.
(iii) There exists a unique $(P, Q) \in \mathsf{Aut}(H_n)$ such that $W_f = W_g Q$.

Proof. $((i) \Rightarrow (ii))$ Let $f \sim g$. Since we have $f(x) = g(xA \oplus \alpha) \oplus \langle x, \beta \rangle \oplus a$, it is not difficult to see that $\zeta_f = \zeta_g P'D$ where $D \in \mathcal{D}_{2^n}$ has the signed sequence of $\langle x, \beta \rangle \oplus a$ as its diagonal and $P' \in \mathcal{S}_{2^n}$ is the permutation matrix corresponding to $xA \oplus \alpha \in \mathsf{AGL}_n$. Then, letting $P = P'D \in \mathcal{S}_{2^n}^{\pm}$ a signed permutation, we obtain $\zeta_f = \zeta_g P$. Uniqueness of the pair (P, Q) follows immediately by the definition of $\mathcal{S}_{2^n}^{\pm}$.

$((ii) \Rightarrow (iii))$ We have $\zeta_f = \zeta_g P$. Then we obtain easily

$$W_f = \zeta_f H_n = \zeta_g P H_n = \zeta_g H_n Q = W_g Q.$$

$((iii) \Rightarrow (i))$ Writing $P = P'D$ for $P \in \mathcal{S}_{2^n}^{\pm}$ with $P' \in \mathcal{S}_{2^n}$ and $D \in \mathcal{D}_{2^n}$, it follows that $W_f = \zeta_f H_n = W_g Q = \zeta_g H_n Q = \zeta_g P H_n$. Hence, this implies $\zeta_f = \zeta_g P = \zeta_g P'D$. By rewriting this equation in the additive case and using the truth tables of the functions f and g, we get $T_f = T_g P' \oplus T_\ell$, where T_ℓ is the corresponding truth table to the diagonal of the matrix D. Interpreting this truth table equation to the algebraic normal forms of the functions $f(x) = g(\tau(x)) \oplus \ell(x)$ is obtained, where $\tau \in \mathcal{S}_{2^n}$ is a permutation of $\mathbb{F}_2^n$ and $\ell \in \mathcal{F}_n$. From Lemma 9.2.3 of [7, p. 102], we conclude that $\tau \in \mathsf{AGL}_n$ and $\ell \in \mathcal{A}_n$ and the result follows. $\square$

We are now ready to prove that we have not only a $1 - 1$ correspondence but also an isomorphy between $\mathsf{AGL}_n \ltimes \mathcal{A}_n$ and $\mathsf{Aut}(H_n)$.

Theorem 2. *The authomorphism group of $2^n \times 2^n$ Sylvester Hadamard matrix is isomorphic to the group of affine equivalence relations acting on n variable Boolean functions, i.e. $\mathsf{Aut}(H_n) \cong \mathsf{AGL}_n \ltimes \mathcal{A}_n$.*

Proof. Bijection between $\mathsf{Aut}(H_n)$ and $\mathsf{AGL}_n \ltimes \mathcal{A}_n$ follows immediately from uniqueness properties of Theorem 1. Consider the map $\varphi : \mathsf{AGL}_n \ltimes \mathcal{A}_n \rightarrow \mathcal{S}_{2^n}^{\pm}$ sending each $(\tau, \ell_{(\beta,a)})$ to $P = P'D$ where $P' \in \mathcal{S}_{2^n}$ is the permutation matrix representing τ and $D \in \mathcal{D}_{2^n}$ is the diagonal matrix having the signed sequence of $\ell_{(\beta,a)}$ as its diagonal. Now, we define the map $\vartheta : \mathsf{AGL}_n \ltimes \mathcal{A}_n \rightarrow \mathsf{Aut}(H_n)$ with $\vartheta : (\tau, \ell_{(\beta,a)}) \mapsto (P, H_n^{-1}PH_n)$. The map ϑ is well-defined since φ is a well-defined map. It certainly is a surjection by the implication $(iii) \Rightarrow (i)$ of Theorem 1. Therefore, we have an explicit version of the bijection given in Theorem 1.

Let $\vartheta((\tau, \ell_{(\beta,a)})) = (P_1, H_n^{-1}P_1H_n)$ and $\vartheta((\tau', \ell_{(\beta',a')})) = (P_2, H_n^{-1}P_2H_n)$. Then we obtain

$$(P_1, H_n^{-1}P_1H_n) \star (P_2, H_n^{-1}P_2H_n) = (P_1 \cdot P_2, H_n^{-1}(P_1 \cdot P_2)H_n).$$

By Definition 1, we let (A, α, β, a) and $(A', \alpha', \beta', a')$ be the representations corresponding to $(\tau, \ell_{(\beta,a)})$ and $(\tau', \ell_{(\beta',a')})$, respectively. Let $P_1 = P_1'D_1$ and $P_2 = P_2'D_2$ with $P_1', P_2' \in \mathcal{S}_{2^n}$ and $D_1, D_2 \in \mathcal{D}_{2^n}$. We have the property that $\varphi((\tau, \ell_{(\beta,a)}) \circ (\tau', \ell_{(\beta',a')}))$ corresponds to a matrix P. Note that by Definition

& Proposition 1, P can be written as $P = P'D$, where $P' = P'_1 P'_2$ is the matrix representing the composition of the permutations determined by $\tau \bullet \tau' : x \mapsto xA'A \oplus \alpha'A \oplus \alpha$ and $D = D_1 P'_1 D_2 (P'_1)^{-1}$ is the diagonal matrix having the signed sequence of $\tau(\ell_{(\beta',a')}) + \ell_{(\beta,a)} : x \mapsto \langle x, \beta' A^t \oplus \beta \rangle \oplus \langle \alpha, \beta' \rangle \oplus a' \oplus a$ as its diagonals. This implies that we have

$$\vartheta((\tau, \ell_{(\beta,a)}) \circ (\tau', \ell_{(\beta',a')})) = (P, H_n^{-1} P H_n)$$
$$= (P'D, H_n^{-1} P' D H_n)$$
$$= (P'_1 P'_2 D_1 P'_1 D_2 (P'_1)^{-1}, H_n^{-1} P'_1 P'_2 D_1 P'_1 D_2 (P'_1)^{-1} H_n)$$
$$= (P_1 \cdot P_2, H_n^{-1}(P_1 \cdot P_2) H_n)$$
$$= (P_1, H_n^{-1} P_1 H_n) \star (P_2, H_n^{-1} P_2 H_n)$$
$$= \vartheta((\tau, \ell_{(\beta,a)})) \star \vartheta((\tau', \ell_{(\beta',a')}))$$

Therefore, it follows $\mathsf{Aut}(H_n) \cong \mathsf{AGL}_n \ltimes \mathcal{A}_n$.   □

Remark 1. There exists another isomorphic group to $\mathsf{Aut}(H_n)$, see Theorem 9.2.4 of [7, p. 103] for details.

The cardinality of $\mathsf{Aut}(H_n)$ can easily be calculated using simple counting arguments and by Theorem 1.

Corollary 1. $|\mathsf{Aut}(H_n)| = |\mathsf{AGL}_n \ltimes \mathcal{A}_n| = 2^{2n+1} \prod_{i=0}^{n-1}(2^n - 2^i)$.

Indeed, based on Theorem 1 it is not difficult to prove the following corollary.

Corollary 2. *Let $f, g \in \mathcal{F}_n$ with $f \sim g$, i.e. there exists $(A, \alpha, \beta, a) \in \mathsf{AGL}_n \ltimes \mathcal{A}_n$ such that $(A, \alpha, \beta, a)g = f$. The corresponding monomial matrix pair $(P, Q) \in \mathsf{Aut}(H_n)$ satisfies the following properties:*

1. $P \in S_{2^n}$ *if and only if $\beta = \alpha_0$ and $a = 0$. Moreover, $Q \in S_{2^n}$ if and only if $\alpha = \alpha_0$.*
2. $P, Q \in D_{2^n}$ *if and only if $A = I_n$ and $\alpha = \alpha_0$.*

4   Nonlinearity Preserving Bijective Mappings

An n-variable Boolean function is an ordered 2^n tuple over $\mathbb{F}_2$. Hence, any bijective mapping acting on such functions can be regarded as a permutation of 2^{2^n} elements. In particular, we can view every bijective mapping as an element of $\mathcal{S}_{2^{2^n}}$.

Let $\psi \in \mathcal{S}_{2^{2^n}}$. We say that ψ preserves nonlinearity if $N_f = N_{\psi f}$ for all $f \in \mathcal{F}_n$. Hence, it follows that ψ preserves nonlinearity if and only if the absolute maximum of the Walsh spectra of f remains invariant, that is

$$\max_{\omega \in \mathbb{F}_2^n} |W_f(\omega)| = \max_{\omega \in \mathbb{F}_2^n} |W_{\psi f}(\omega)| \quad \text{for all } f \in \mathcal{F}_n.$$

We denote the set of all nonlinearity preserving bijective maps acting on the n-variables Boolean functions by

$$\mathcal{P}_n(N) = \{\psi \in \mathcal{S}_{2^{2^n}} \mid N_f = N_{\psi f} \text{ for all } f \in \mathcal{F}_n\}.$$

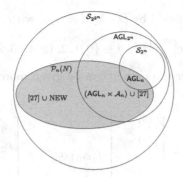

Fig. 1. Current state of nonlinearity preserving bijective transformations. (Here, NEW represents new maps coming from the results of this work.)

Proposition 3. $\mathcal{P}_n(N)$ *is a subgroup of* $\mathcal{S}_{2^{2^n}}$.

Proof. It is clear that identity mapping of $\mathcal{S}_{2^{2^n}}$ lies in $\mathcal{P}_n(N)$. We take any $\psi, \theta \in \mathcal{P}_n(N)$. We have $N_f = N_{\psi f}$ and $N_f = N_{\theta f}$ for all $f \in \mathcal{F}_n$. Then, it follows by Definition & Proposition 3 that $N_f = N_{\theta(\psi f)} = N_{(\theta \circ \psi)f}$. $\square$

In the previous section, we revisited the action of $\mathsf{AGL}_n \ltimes \mathcal{A}_n$ on $\mathcal{F}_n$. Proposition 2 implies that nonlinearity is preserved under the action of $\mathsf{AGL}_n \ltimes \mathcal{A}_n$. We further notice that $\mathsf{AGL}_n \ltimes \mathcal{A}_n$ forms naturally a subgroup of $\mathcal{P}_n(N)$.

Theorem 1 implies that $T_f = T_g P' \oplus T_{\ell_{(\beta,a)}}$ holds for any $f, g \in \mathcal{F}_n$ with $(A, \alpha, \beta, a)g = f$, where P' is the permutation matrix representation of the map $x \mapsto xA + \alpha$. Examining the action of $\mathsf{AGL}_{2^n} := \mathsf{GL}_{2^n} \ltimes \mathbb{F}_2^{2^n}$ on $\mathcal{F}_n$ gives us a natural way of classifying such bijective mappings. For any $(A, h) \in \mathsf{AGL}_{2^n}$, this action can naturally be made explicit as follows

$$\tau f = (A, h)f := T_f A \oplus T_h.$$

It is not difficult to see that $\mathsf{AGL}_n \ltimes \mathcal{A}_n \subseteq \mathsf{AGL}_{2^n}$. Furthermore, AGL_{2^n} can be regarded as a subgroup of $\mathcal{S}_{2^{2^n}}$. These two observations led not only to generalize the notion of affine equivalence to an action under the larger group AGL_{2^n} (instead of considering the equivalence under $\mathsf{AGL}_n \ltimes \mathcal{A}_n$), but also to prove the existence of new classes of nonlinearity preserving mappings lying in $\mathsf{AGL}_{2^n} \setminus \mathsf{AGL}_n \ltimes \mathcal{A}_n$ [27].

Any bijective map $\psi \in \mathcal{S}_{2^{2^n}}$ acting on $\mathcal{F}_n$ can be viewed as a vectorial Boolean function $\psi : \mathbb{F}_2^{2^n} \to \mathbb{F}_2^{2^n}$. More explicitly, the image of an element $x = (x_0, x_1, \ldots, x_{2^n-1}) \in \mathbb{F}_2^{2^n}$ is given by

$$\psi(x_0, x_1, \ldots, x_{2^n-1}) = (f^0(x_0, x_1, \ldots, x_{2^n-1}), \ldots, f^{2^n-1}(x_0, x_1, \ldots, x_{2^n-1})),$$

where each *coordinate function* f^i is an element of $\mathcal{F}_n$ with $0 \leq i \leq 2^n - 1$.

Since each $f^i \in \mathcal{F}_{2^n}$ can be represented by its unique algebraic normal form $f^i(x_0, x_2, \cdots, x_{2^n}) = c_0^{(i)} \oplus c_1^{(i)} x_0 \oplus \ldots \oplus c_{12\ldots 2^n}^{(i)} x_1 x_2 \cdots x_{2^n}$ we have

$$\psi : T_f \mapsto (\lambda_0 \oplus A^t T_f^t \oplus \lambda_{12} f(\alpha_0) f(\alpha_1) \oplus \cdots \oplus \lambda_{12\ldots 2^n} f(\alpha_0) f(\alpha_1) \cdots f(\alpha_{2^n-1}))^t,$$

where A^t is the matrix defined by $[\lambda_1 \ \lambda_2 \ \ldots \ \lambda_{2^n}]$ with

$$\lambda_j = (c_j^{(0)}, c_j^{(1)}, \ldots, c_j^{(2^n-1)})^t, \ j \in \{0, 1, 2, 12, 3, 13, 23, \ldots, 123\cdots 2^n\}.$$

The image of T_f under the map ψ can be given more explicitly as follows

$$\psi : T_f \longmapsto \left(\underbrace{\begin{bmatrix} c_0^{(0)} \\ c_0^{(1)} \\ \vdots \\ c_0^{(2^n-1)} \end{bmatrix}}_{\lambda_0} \oplus \underbrace{\begin{bmatrix} c_1^{(0)} \\ c_1^{(1)} \\ \vdots \\ c_1^{(2^n-1)} \end{bmatrix}}_{\lambda_1} f(\alpha_0) \oplus \cdots \oplus \underbrace{\begin{bmatrix} c_{2^n}^{(0)} \\ c_{2^n}^{(1)} \\ \vdots \\ c_{2^n}^{(2^n-1)} \end{bmatrix}}_{\lambda_{2^n}} f(\alpha_{2^n-1}) \oplus \right.$$

$$\left. \underbrace{\begin{bmatrix} c_{12}^{(0)} \\ c_{12}^{(1)} \\ \vdots \\ c_{12}^{(2^n-1)} \end{bmatrix}}_{\lambda_{12}} f(\alpha_0) f(\alpha_1) \oplus \cdots \oplus \underbrace{\begin{bmatrix} c_{12\cdots 2^n}^{(0)} \\ c_{12\cdots 2^n}^{(1)} \\ \vdots \\ c_{12\cdots 2^n}^{(2^n-1)} \end{bmatrix}}_{\lambda_{12\cdots 2^n}} f(\alpha_0) \cdots f(\alpha_{2^n-1}) \right)^t.$$

The representation given above can be explained as follows

- $\psi \in \mathsf{AGL}_n \ltimes \mathcal{A}_n$ if
  - $\lambda_0 \in \mathcal{A}_n$ and $A^t \in \mathcal{S}_{2^n}$ correspond to the matrix representation of an element in AGL_n and
  - $\lambda_j = (0 \ 0 \ \ldots \ 0)^t$ for all $j \neq 0, 1, 2, 3, 4, \ldots, 2^n$.
- $\psi \in \mathsf{AGL}_{2^n}$ if
  - $\lambda_0 \in \mathbb{F}_2^{2^n}$ and $A \in \mathsf{GL}_{2^n}$ and
  - $\lambda_j = (0 \ 0 \ \ldots \ 0)^t$ for all $j \neq 0, 1, 2, 3, 4, \ldots, 2^n$.
- $\psi \in \mathcal{S}_{2^{2n}}$ is called *non-affine* if it has at least one non-zero λ_j, for $j \in \{12, 13, \ldots, 12\cdots 2^n\}$.

In [27, Proposition 3.8], the existence of nonlinearity preserving non-affine bijective mappings is proved by explicit construction. Here, we prove that (even if they are non-affine mappings) these mappings lie actually in $\mathsf{AGL}_n \ltimes \mathcal{A}_n$ for a fixed function.

Proposition 4. *Let the notations be as above. Let further $\psi \in \mathcal{S}_{2^{2n}}$ be a non-affine mapping satisfying*

1. *$\lambda_0 \in \mathcal{A}_n$,*
2. *The matrix $A = (P' \oplus B)$ with $P' \in \mathcal{S}_{2^n}$ corresponds to a matrix representation of an element in AGL_n and $B = [\varepsilon_1 \ \varepsilon_2 \ \ldots \ \varepsilon_{2^n}]$ with $\varepsilon_i \in \mathcal{A}_n$, $1 \leq i \leq 2^n$,*
3. *$\lambda_j \in \mathcal{A}_n$ for all $j \in \{12, 13, \ldots, 12\cdots 2^n\}$ where at least one of $\lambda_j \neq (0\ldots 0)^t$.*

Then ψ is an element of $\mathsf{AGL}_n \ltimes \mathcal{A}_n$ for a fixed function $f \in \mathcal{F}_n$.

Proof. Since $f \in \mathcal{F}_n$ is fixed

$$T_f \mapsto ((P' \oplus B)^t T_f^t \oplus \lambda_0 \oplus \lambda_{12} f(\alpha_0) f(\alpha_1) \oplus \cdots \oplus \lambda_{12 \cdots 2^n} f(\alpha_0) f(\alpha_1) \cdots f(\alpha_{2^n-1}))^t$$

reduces to the mapping $T_f \mapsto T_f P' \oplus T_{\ell_{(\beta,a)}}$, where $T_{\ell_{(\beta,a)}}$ is strictly determined by summation of ε_i and λ_j based on $Supp(f)$. Note that $T_{\ell_{(\beta,a)}}$ is nothing but the truth table of an affine function.   □

A computer experiment, in [27], shows that $\mathcal{P}_2(N)$ consists of only the mappings given in Proposition 4. For $n \geq 3$, we prove now the existence of new non-affine mappings not belonging to the class of Proposition 4 using basic counting techniques and nonlinearity distributions.

Theorem 3. *For $n \geq 3$ there exist nonlinearity preserving non-affine mappings not lying in the class of Proposition 4.*

Proof. For $n = 3$ we have 2^{64} choices for the matrix A and 16 choices for each λ_i for $i \in \{0, 12, 13, \ldots, 12 \cdots 2^n\}$. It means that the number of bijective mappings given in Proposition 4 is strictly less than $2^{1056} = 2^{64} \times 16^{248}$. Similarly, for $n = 4$, the number of bijective mappings is less than 2^{327856}. Note that these cardinalities are strictly less than the values of $|\mathcal{P}_3(N)|$ (resp. $|\mathcal{P}_4(N)|$). See Table 1 for the illustration of $n \leq 6$. Therefore, for $n = 3, 4$ Proposition 4 type mappings do not cover all of the nonlinearity preserving mappings.

For $n \geq 5$ the ratio of $|\mathcal{A}_n|$ to $|\mathcal{F}_n|$ will decrease as n increases. Thus, for $n \geq 3$, mappings coming from Proposition 4 constitute just a proper subset of $\mathcal{P}_n(N)$.   □

Even if these new mappings have not been described with their algebraic normal form, we present some examples for $n = 3, 4, 6$ in Appendix A to illustrate the situation.

Exact classification of $\mathcal{P}_n(N)$ is still an open problem. The current state of the classification of nonlinearity preserving mappings is illustrated in Fig. 1. However, for values $n \leq 6$, where nonlinearity distributions are easy to calculate, the cardinality of $\mathcal{P}_n(N)$ can also be computed. We present for $n \leq 6$ the cardinality of $\mathcal{P}_n(N)$ in Table 1, where the computation is based on the nonlinearity distribution given in Appendix B.

Table 1. $|\mathcal{P}_n(N)|$ and $|\mathcal{S}_{2^{2n}}|$ values for $n \leq 6$

n	$|\mathcal{P}_n(N)|$	$|\mathcal{S}_{2^{2n}}| = 2^{2^n}!$
2	$8! \times 8! \approx 10^{11}$	$2^{2^2}! \approx 10^{13}$
3	$16! \times 128! \times 112! \approx 10^{413}$	$2^{2^3}! \approx 10^{506}$
4	$32! \times 512! \times \cdots \times 896! \approx 10^{249727}$	$2^{2^4}! \approx 10^{287194}$
5	$64! \times \cdots \times 27387136! \approx 10^{3,66x10^{10}}$	$2^{2^5}! \approx 10^{3.95x10^{10}}$
6	$128! \times \cdots \times 5425430528! \approx 10^{3.32x10^{20}}$	$2^{2^6}! \approx 10^{3,47x10^{20}}$

5   Automorphism Group of Nonlinearity Classes

There are some proposals in the literature stating that the automorphism group of $\mathcal{B}_n$ is just the group $\mathsf{AGL}_n \ltimes \mathcal{A}_n$ [5,32]. Definitely each element of $\mathsf{AGL}_n \ltimes \mathcal{A}_n$ gives us an automorphism in the sense of stabilizing the bentness property of the functions. However, there are other transformations mapping $\mathcal{B}_n$ to itself which are not lying in the group $\mathsf{AGL}_n \ltimes \mathcal{A}_n$. This observation suggests that the definition of automorphism group of bent functions should be reformulated. For example, for $n = 4$, there are $|\mathcal{P}_4(N)| \approx 2^{829564}$ bijective mappings preserving nonlinearity. However, only $896! \approx 2^{7500}$ of them constitute different permutations on $\mathcal{B}_4$, while $|\mathsf{AGL}_4 \ltimes \mathcal{A}_4| \approx 2^{23}$. Thus, $\mathsf{AGL}_4 \ltimes \mathcal{A}_4$ must only be a proper subset of the whole automorphism group of bent functions.

This new treatment of automorphisms suggests that main cryptographical design concerns for the automorphism groups in the common literature should be criticized. For example in [32], it is shown that $\mathsf{AGL}_n \ltimes \mathcal{A}_n$ preserves nonlinearity by preserving the distance between the function to its closest affine function. However, it is an unnecessarily strong condition for cryptographic purposes because preserving the distance of any function to the affine function class is ample. More generally, determination of the automorphism group of nonlinearity classes should be studied as a subgroup of $\mathcal{S}_{2^{2n}}$ since $\mathsf{AGL}_n \ltimes \mathcal{A}_n \leq \mathcal{P}_n(N)$. To be more accurate, instead of mentioning about the automorphism group shortly, one may at least define the considered group that acts on Boolean functions explicitly.

6   Conclusion

Studying the elements $\mathcal{S}_{2^{2n}}$ and classifying them with respect to nonlinearity preserving property is still an open problem. Despite the fact that such a research may seem to be highly involved due to the huge cardinality of the mappings, it may lead to a deeper insight to the nonlinear functions or nonlinearity classes. Formerly, it was proposed that automorphism group of bent functions is $\mathsf{AGL}_n \ltimes \mathcal{A}_n$. In this work, it is shown that there are new transformations keeping nonlinearity invariant. Therefore, such propositions should be reexamined by means of considering a larger set than $\mathsf{AGL}_n \ltimes \mathcal{A}_n$. For future studies, it would be interesting to further investigate nonlinearity preserving mappings in order to construct those mappings with explicit methods.

Acknowledgements. We would like to thank the anonymous reviewers for their valuable comments on the preliminary version of this work. Third author is partly supported by a joint research project funded by Bundesministerium für Bildung und Forschung (BMBF), Germany (01DL12038) and TÜBİTAK Turkey (TBAG-112T011).

Appendix A: Examples of New Transformations for $n = 3, 4, 6$

Due to the space constraints the algebraic normal form of the nonlinearity preserving transformations are not given explicitly for $n \geq 4$. However, since any transformation is an element of $S_{2^{2n}}$, it is possible to represent the permutations as a product of disjoint cycles. Within these cycles, the functions will be given by their truth table in hexadecimal value without the "0x" prefix.

Example 1. Let $\psi \in S_{2^{2^3}}$ be,

$$\psi : T_f \mapsto \big(\lambda_0 \oplus A^t T_f^t \oplus \lambda_{123457} f(\alpha_0) f(\alpha_1) f(\alpha_2) f(\alpha_3) f(\alpha_4) f(\alpha_6) \oplus$$
$$\lambda_{1234578} f(\alpha_0) f(\alpha_1) f(\alpha_2) f(\alpha_3) f(\alpha_4) f(\alpha_6) f(\alpha_7) \oplus$$
$$\lambda_{123456} f(\alpha_0) f(\alpha_1) f(\alpha_2) f(\alpha_3) f(\alpha_4) f(\alpha_5) \oplus$$
$$\lambda_{1234568} f(\alpha_0) f(\alpha_1) f(\alpha_2) f(\alpha_3) f(\alpha_4) f(\alpha_5) f(\alpha_7)\big)^t$$

where $\lambda_0 = [00001111]^t$, $\lambda_{123457} = \lambda_{1234578} = \lambda_{123456} = \lambda_{1234568} = [00010100]^t$ and A^t is the matrix;

$$\begin{bmatrix} 1 & 0 & 0 & 0 & 0 & 0 & 0 & 0 \\ 0 & 0 & 0 & 0 & 1 & 0 & 0 & 0 \\ 0 & 0 & 1 & 0 & 0 & 0 & 0 & 0 \\ 0 & 0 & 0 & 0 & 0 & 0 & 1 & 0 \\ 1 & 0 & 1 & 1 & 1 & 1 & 1 & 1 \\ 1 & 1 & 1 & 1 & 1 & 0 & 1 & 1 \\ 1 & 1 & 1 & 0 & 1 & 1 & 1 & 1 \\ 1 & 1 & 1 & 1 & 1 & 1 & 1 & 0 \end{bmatrix}.$$

Trivially, ψ is not an affine mapping. In particular, it does not satisfy the conditions of Proposition 4 since $(0, 0, 0, 1, 0, 1, 0, 0)$ is not a truth table of an affine function. Moreover, it can be easily checked that this map is invertible and preserves nonlinearity for all functions.

Example 2. Assume $\psi \in S_{2^{2^4}}$ be a permutation of $\mathbb{F}_2^{2^4}$ with cycle representation,
(0000, 6699, f0f0, a9e2, 9999, 9696, 5aa5, 3cc3) (0081, 00bd, 01f7) (0107, 0c 3f, f0ff, 0a41, 5fcc, bf1f, 2e8b, 0251, 30cf, 13d3) (0471, fee4, 0495, 242f, 6c 7f) (065c, 8d28, 0a9c, 4478, 1d7b, 2eed, de74) (09b1, 2882, 60e8, 1284, e44e) (12ed, 1441, 1482, 1428) (1414, 1698, 16c2) (5a58, 7996, 5a5b).

It can be easily verified that ψ keeps nonlinearity values invariant for all $f \in \mathcal{F}_4$, that is $\psi \in \mathcal{P}_4(N)$. When the algebraic normal form of ψ is written explicitly, it will be seen that some non-affine terms are not the truth table of an affine function. Thus ψ is not of the form given in Proposition 4.

Example 3. Let $\psi \in S_{2^{2^6}}$ be the permutation such that its disjoint cycle representation is
(0217177f68818115, 051307ff58900943)
(c003033f6aa9a995, c113077f6889a115, c113077f984a9215).

Trivially, ψ maps all functions to itself except the ones in the disjoint cycles, i.e. it keeps nonlinearity invariant for those functions. The first cycle has functions whose nonlinearity values are 26 whereas the second has 28. Thus, $\psi \in \mathcal{P}_6(N)$. However, the bent function "c003033f6aa9a995" has algebraic degree 2 while its image under ψ "c113077f6889a115" has algebraic degree 3. ψ can not be of the form as given in Proposition 4 since in that case it should have also keep the algebraic degree invariant.

Appendix B: Nonlinearity Classes for $n \leq 6$

Table 2. Nonlinearity class cardinalities for $n \leq 6$

N_f	$n = 2$	$n = 3$	$n = 4$	$n = 5$	$n = 6$
0	8	16	32	64	128
1	8	128	512	2048	8192
2	-	112	3840	31744	258048
3	-	-	17920	317440	5332992
4	-	-	28000	2301440	81328128
5	-	-	14336	12888064	975937536
6	-	-	896	57996288	9596719104
7	-	-	-	215414784	79515672576
8	-	-	-	647666880	566549167104
9	-	-	-	1362452480	3525194817536
10	-	-	-	1412100096	19388571496448
11	-	-	-	556408832	95180260073472
12	-	-	-	27387136	420379481991168
13	-	-	-	-	1681517927964672
14	-	-	-	-	6125529594728448
15	-	-	-	-	20418431982428160
16	-	-	-	-	62526600834171264
17	-	-	-	-	176395152249028608
18	-	-	-	-	458313050588725248
19	-	-	-	-	1087405010755682304
20	-	-	-	-	2291582136636334080
21	-	-	-	-	4011570131804454912
22	-	-	-	-	5097726702198767616
23	-	-	-	-	3821934098435833856
24	-	-	-	-	1305039828998603264
25	-	-	-	-	103868560519987200
26	-	-	-	-	1617838297055232
27	-	-	-	-	347227553792
28	-	-	-	-	5425430528

References

1. Braeken, A., Nikova, S., Borissov, Y.: Classification of cubic boolean functions in 7 variables. In: Proceedings of 26th Symposium on Information Theory in the Benelux, Brussels, Belgium, pp. 285–292 (2005)
2. Braeken, A., Borissov, Y., Nikova, S., Preneel, B.: Classification of boolean functions of 6 variables or less with respect to some cryptographic properties. In: Caires, L., Italiano, G.F., Monteiro, L., Palamidessi, C., Yung, M. (eds.) ICALP 2005. LNCS, vol. 3580, pp. 324–334. Springer, Heidelberg (2005)
3. Budaghyan, L., Carlet, C.: CCZ-equivalence and Boolean functions. Cryptology ePrint Archive, Report 2009/063 (2009)
4. Carlet, C., Charpin, P., Zinoviev, V.: Codes, bent functions and permutations suitable for DES-like cryptosystems. Des. Codes Cryptgr. **15**, 125–156 (1998)
5. Carlet, C., Mesnager, S.: On Dillon's class H of bent functions, Niho bent functions and O-Polynomials. Cryptology ePrint Archive, Report 2010/567 (2010)
6. Crama, Y., Hammer, P.L.: Boolean Models and Methods in Mathematics, Computer Science, and Engineering. Encyclopedia of Mathematics and its Applications, vol. 134. Cambridge University Press, Cambridge (2010)
7. De Launey, W., Flannery, D.L.: Algebraic Design Theory, vol. 175. American Mathematical Society, Providence (2011)
8. Denev, J., Tonchev, V.: On the number of equivalence classes of Boolean functions under a transformation group (Corresp.). IEEE Trans. Inf. Theory **26**(5), 625–626 (1980)
9. Dillon, J.: A survey of bent functions. NSA Tech. J., 191–215 (1972). Special Issue
10. Fuller, J.E.: Analysis of affine equivalent Boolean functions for cryptography. Ph.D. thesis, Queensland University of Technology, Queensland, Australia (2003)
11. Hadamard, J.: Résolution d'une question relative aux déterminants. Bull. Sci. Math. **2**(17), 240–246 (1893)
12. Hall Jr., M.: Note on the Mathieu group $\mathcal{M}_{\infty \in}$. Arch. Math. **13**, 334–340 (1962)
13. Harrison, M.A.: The number of classes of invertible Boolean functions. J. ACM (JACM) **10**(1), 25–28 (1963)
14. Harrison, M.A.: On the classification of Boolean functions by the general linear and affine groups. J. Soc. Ind. Appl. Math. **12**(2), 285–299 (1964)
15. Hou, X.D.: $AGL(m, 2)$ acting on $R(r, m)/R(s, m)$. J. Algebra **171**(3), 921–938 (1995)
16. Langevin, P., Leander, G.: Counting all bent functions in dimension eight 99270589265934370305785861242880. Des. Codes Cryptgr. **59**(1–3), 193–205 (2010)
17. Leon, J.S.: An algorithm for computing the automorphism group of a Hadamard matrix. J. Comb. Theory Ser. A **27**(3), 289–306 (1979)
18. MacWilliams, F.J., Sloane, N.J.A.: The Theory of Error-Correcting Codes, 2nd edn. North-holland Publishing Company, Amsterdam (1978)
19. Maiorana, J.A.: A classification of the cosets of the Reed-Muller code $R(1, 6)$. Math. Comput. **57**(195), 403–414 (1991)
20. Meier, W., Staffelbach, O.: Nonlinearity criteria for cryptographic functions. In: Quisquater, J.-J., Vandewalle, J. (eds.) EUROCRYPT 1989. LNCS, vol. 434, pp. 549–562. Springer, Heidelberg (1990)
21. Pieprzyk, J., Finkelstein, G.: Towards effective nonlinear cryptosystem design. IEE Proc. E Comput. Digit. Tech. **135**(6), 325–335 (1988)

22. Preneel, B.: Analysis and design of cryptographic hash functions. Ph.D. thesis, Katholieke Universiteit Leuven, Leuven, Belgium, January 1993
23. Rothaus, O.S.: On "bent" functions. J. Comb. Theory Ser. A **20**(3), 300–305 (1976)
24. Seberry, J., Zhang, X.-M.: Highly nonlinear 0–1 balanced Boolean functions satisfying strict avalanche criterion (extended abstract). In: Zheng, Y., Seberry, J. (eds.) AUSCRYPT 1992. LNCS, vol. 718, pp. 143–155. Springer, Heidelberg (1993)
25. Sertkaya, İ.: Nonlinearity preserving post-transformations. Master's thesis, Middle East Technical University, Ankara, Turkey, June 2004
26. Sertkaya, İ., Doğanaksoy, A.: On nonlinearity preserving bijective transformations. In: Proceedings of the National Cryptology Symposium II, Ankara, Turkey, pp. 27–36 (2006)
27. Sertkaya, İ., Doğanaksoy, A.: Some results on nonlinearity preserving bijective transformations. In: Proceedings of BFCA 2007 Conference, Paris, France, pp. 27–42 (2007)
28. Sertkaya, İ., Doğanaksoy, A.: On the affine equivalence and nonlinearity preserving bijective mappings. Cryptology ePrint Archive, Report 2010/655 (2010)
29. Stone, H.S., Jackson, C.L.: Structures of the affine families of switching functions. IEEE Trans. Comput. **3**, 251–257 (1969)
30. Strazdins, I.: Universal affine classification of Boolean functions. Acta Applicandae Mathematica **46**(2), 147–167 (1997)
31. Sylvester, J.J.: Thoughts on inverse orthogonal matrices, simultaneous sign successions, and tessellated pavements in two or more colors, with applications to Newton's rule, ornamental tile-work, and the theory of numbers. Philos. Mag. **34**, 461–475 (1867)
32. Tokareva, N.: Automorphism group of the set of all bent functions. Cryptology ePrint Archive, Report 2010/255 (2010)

On Verification of Restricted Extended Affine Equivalence of Vectorial Boolean Functions

Ferruh Özbudak[1,2], Ahmet Sınak[2,4], and Oğuz Yayla[2,3(✉)]

[1] Department of Mathematics, Middle East Technical University,
Dumlupınar Bul., No:1, 06800 Ankara, Turkey
ozbudak@metu.edu.tr
[2] Institute of Applied Mathematics, Middle East Technical University,
Dumlupınar Bul., No:1, 06800 Ankara, Turkey
[3] Johann Radon Institute for Computational and Applied Mathematics, Austrian
Academy of Sciences, Altenberger Strasse 69, 4040 Linz, Austria
oguz.yayla@gmail.com
[4] Department of Mathematics and Computer Sciences, Necmettin Erbakan
University, Meram Yerleşkesi, 42090 Konya, Turkey
ahmet.sinak@metu.edu.tr

Abstract. Vectorial Boolean functions are used as substitution boxes in cryptosystems. Designing inequivalent functions resistant to known attacks is one of the challenges in cryptography. In doing this, finding a fast technique for determining whether two given functions are equivalent is a significant problem. A special class of the equivalence called restricted extended affine (REA) equivalence is studied in this paper. We update the verification procedures of the REA-equivalence types given in the recent work of Budaghyan and Kazymyrov (2012). In particular, we solve the system of linear equations simultaneously in the verification procedures to get better complexity. We also present the explicit number of operations of the verification procedures of these REA-equivalence types. Moreover, we construct two new REA-equivalence types and present the verification procedures of these types with their complexities.

Keywords: Vectorial Boolean functions · EA-equivalence · REA-equivalence

1 Introduction

The verification and generation of vectorial Boolean functions with optimal characteristics, resistant to known attacks, are two important problems in cryptography. The purpose of this paper is to contribute to solving the verification problem of vectorial Boolean functions. Sometimes equivalence classes of vectorial Boolean functions are helpful for cryptographers to achieve necessary properties without losing others (i.e. δ-uniformity, nonlinearity, etc.). When a new vectorial Boolean function is proposed, it is necessary to verify whether it is equivalent to already known ones [6]. The complexity of exhaustive search for

© Springer International Publishing Switzerland 2015
Ç. Koç et al. (Eds.): WAIFI 2014, LNCS 9061, pp. 137–154, 2015.
DOI: 10.1007/978-3-319-16277-5_8

checking extended affine (EA)-equivalence of two functions from $\mathbb{F}_2^n$ to itself equals $O(2^{3n^2+2n})$ [3]. When $n = 6$, the complexity is 2^{120} which is difficult to accomplish with the existing resources and computers. Thus, *restricted extended affine (REA)-equivalence* was studied in [1] and [3]. A. Biryukov et al. [1] showed that if given functions are permutations of $\mathbb{F}_2^n$, the complexity of determining REA-equivalence equals $O(n^2 2^n)$ for the case of linear equivalence, and $O(n^2 2^{2n})$ for affine equivalence. In addition, Budaghyan and Kazymyrov [3] showed that more four types of REA-equivalence can be verified for arbitrary functions. The types given in [3] and two new types of REA-equivalence are studied in this paper.

Our contributions: We firstly update the verification procedures of REA-equivalence types namely types I to IV given in [3]. Precisely, we apply Gaussian elimination method simultaneously for solving the system of equations in the verification procedure to improve the complexities of types I and IV. In addition to that, we show that only finding the solution of the system of equations corresponding to the REA-equivalence types is not enough. One also needs to check whether the solution satisfies the REA-equivalence relation. Thus we include this check operation into the verification procedures. Our second contribution is that we obtain two new REA-equivalence types and analyze their verification procedures, namely types V and VI. We summarize our results in Tables 2, 4, 6 and 7.

Overview of the paper: The paper is organized as follows. Section 2 introduces basic definitions, equivalence algorithms and their properties. Section 3 presents previous results mainly given in [3]. We also give the complexity of solving the system of equations simultaneously by the Gaussian elimination method in Sect. 3. Section 4 provides verification procedures of REA-equivalence types with their complexities. Section 5 gives a brief summary of this paper.

2   Preliminaries

In this section, we present fundamental definitions and results used in the subsequent sections. For further explanations, definitions, applications and previous work, please see [1,4–6,8] and references there in. Throughout this paper, the field addition modulo 2 (in $\mathbb{F}_2$) will be denoted by *XOR*, and the field multiplication modulo 2 (in $\mathbb{F}_2$) will be denoted by *AND*. However, the addition of the elements of the finite fields $\mathbb{F}_{2^n}$ is denoted by $+$ despite the fact that it is performed in characteristic 2. Since $\mathbb{F}_2^n$ can be identified with $\mathbb{F}_{2^n}$, we shall also denote by $+$ the addition of vectors of $\mathbb{F}_2^n$ when $n > 1$.

Definition 1. *Let n and m be two positive integers. The functions from $\mathbb{F}_2^n$ to $\mathbb{F}_2^m$ are called (n, m)-functions. Such function F being given, the Boolean functions $f_1, \ldots, f_m$ defined at every $x \in \mathbb{F}_2^n$, by $F(x) = (f_1(x), \ldots, f_m(x))$, are called the coordinate functions of F. When the numbers m and n are not specified, (n, m)-functions are called multi-output Boolean functions or vectorial Boolean functions or S-boxes.*

Finding a vectorial Boolean function over a finite field satisfying some crypto-graphic criteria is one of the main problems. Once a new function is found, the second problem occurs in determining whether it is equivalent or not in some sense to any of the functions already known, see [6] for details. We will deal with the later one in this paper. Some definitions of equivalence relations for vectorial Boolean functions will be given in Sect. 2.1.

2.1 Equivalences

Equivalence classes provide a powerful tool for dividing the vectorial Boolean functions into classes in which they preserve some cryptographic properties. More studies are needed to fully characterize the equivalence classes of vecto-rial Boolean functions with respect to their cryptographic properties, as well as to determine the number of equivalence classes for functions with more than six input variables. Another important problem in this area is to find a fast technique for determining whether two functions with more than six input vari-ables are equivalent or not. Several techniques exist, but the most general known equivalences appear to be *the Carlet-Charpin-Zinoviev (CCZ)-equivalence* and *the EA-equivalence*. The CCZ-equivalence (see [4,5]) partitions the set of func-tions into classes with the same nonlinearity and differential uniformity. Thus, the resistance of a function to *linear and differential attacks* is CCZ-invariant. CCZ-equivalent functions have also the same weakness/strength with respect to algebraic attacks. Now, we introduce the EA-equivalence for two vectorial Boolean functions.

Definition 2. *[3] Let $F, G : \mathbb{F}_{2^n} \to \mathbb{F}_{2^m}$ be vectorial Boolean functions. The functions F, G are called extended affine equivalent (EA-equivalent) if there exist an affine permutation A_1 of $\mathbb{F}_{2^m}$, an affine permutation A_2 of $\mathbb{F}_{2^n}$ and a linear function L_3 from $\mathbb{F}_{2^n}$ to $\mathbb{F}_{2^m}$ such that*

$$F(x) = A_1 \circ G \circ A_2(x) + L_3(x).$$

For $A_1(x) = M_1 \cdot x + V_1$, $A_2(x) = M_2 \cdot x + V_2$ and $L_3(x) = M_3 \cdot x$ where $M_1 \in \mathbb{F}_2^{m \times m}$ and $M_2 \in \mathbb{F}_2^{n \times n}$ are non-singular matrices, $V_1 \in \mathbb{F}_2^{m \times 1}$, $V_2 \in \mathbb{F}_2^{n \times 1}$, and $M_3 \in \mathbb{F}_2^{m \times n}$, the EA-equivalence is represented in the matrix form as follows;

$$F(x) = M_1 \cdot G(M_2 \cdot x + V_2) + M_3 \cdot x + V_1.$$

It is obvious that the complexity of verification of *EA-equivalence* is too high. The restricted form of EA-equivalence, called REA-equivalence, is defined in [3] as follows.

Definition 3. *Let $F, G : \mathbb{F}_{2^n} \to \mathbb{F}_{2^m}$ be vectorial Boolean functions. The func-tions F, G are called restricted EA-equivalent (REA-equivalent) if at least one of the following holds*

- M_1 or/and M_2 are the identity matrices $\mathcal{I}$,
- M_3 is the zero matrix or the identity matrix $\mathcal{I}$,
- V_1 or/and V_2 are the zero vectors.

Linear equivalence (LE) and *affine equivalence (AE)* (see [1]) are particular cases of the REA-equivalence. Some REA-equivalence algorithms are presented in Table 1.

Table 1. Some types of REA-equivalence

M_1	M_2	M_3	V_1	V_2	Types	REA-equivalence	Source
		0	0	0	LE	$F(x) = M_1 \cdot G(M_2 \cdot x)$	[1]
		0			AE	$F(x) = M_1 \cdot G(M_2 \cdot x + V_2) + V_1$	[1]
	$\mathcal{I}$	0		0	I	$F(x) = M_1 \cdot G(x) + V_1$	[3]
$\mathcal{I}$		0	0		II	$F(x) = G(M_2 \cdot x + V_2)$	[3]
$\mathcal{I}$	$\mathcal{I}$			0	III	$F(x) = G(x) + M_3 \cdot x + V_1$	[3]
	$\mathcal{I}$			0	IV	$F(x) = M_1 \cdot G(x) + M_3 \cdot x + V_1$	[3]
$\mathcal{I}$				0	V	$F(x) = G(M_2 \cdot x) + M_3 \cdot x + V_1$	New
$\mathcal{I}$		0		0	VI	$F(x) = G(M_2 \cdot x) + V_1$	New

Notice that EA-equivalent functions are CCZ-equivalent, but CCZ-equivalent functions are not necessarily EA-equivalent even if these functions are inverse of each other [4]. The resistance of a function to linear and differential attacks is EA-invariant since EA-equivalent functions have the same differential uniformity and the same nonlinearity. Therefore, EA-equivalence classes of functions over finite fields preserve APN property (*differentially 2-uniform*) and AB property (*non-linearity*) desirable for S-boxes [4]. They lead to infinite classes of AB and APN polynomials, which are EA-inequivalent to some power functions. The algebraic degree of a function (if it is not affine) is invariant under the EA-equivalence. It is obvious that the properties which are EA-invariant is also REA-invariant.

3   Previous Result

In this section, the previous results [3] related to solving the system of equations is presented. Let $F : \mathbb{F}_2^n \rightarrow \mathbb{F}_2^m$ and $G : \mathbb{F}_2^n \rightarrow \mathbb{F}_2^m$ be two vectorial Boolean functions. The aim is to find an $m \times m$ matrix M satisfying the equation

$$F(x) = M \cdot G(x). \tag{1}$$

The trivial method to find M is the exhaustive search whose complexity is $O(2^{m^2+n})$ [3]. The other method is the natural method which is based on solving the system of 2^n equations with m variables. Budaghyan and Kazymyrov [3, Proposition 2] use the Williams method [10] to reach that this system can be solved with the complexity of $O(\max\{2^n, m\}^2) = O(2^{2n})$. It is obvious that one row of M can be found by solving one system of equations. It is required to solve m-systems of 2^n equations in m unknowns to find the matrix M. Consequently, the complexity of finding M (as given in [3]) equals

$$O(m \cdot \max\{2^n, m\}^2) = O(m2^{2n}). \tag{2}$$

Apparently, this complexity is extremely high. To reduce it, we propose to use *Gaussian elimination method* simultaneously to solve the system of Eq. (1). As solving the system of Eq. (1) requires the elimination of the same matrix for extracting each row of M, this method has obvious advantageous. We present the detailed explanations and proofs in [9, Chapter 4]. See also the proof of Proposition 1 below.

Lemma 1. *The complexity of solving the m-systems of 2^n equations in m unknowns, with simultaneous Gaussian elimination method is*

$$(\frac{3m^2 + m}{2})(2^n - 1) \tag{3}$$

field additions modulo 2, which gives the complexity of finding $m \times m$ matrix M in the Eq. (1).

Simultaneous application of the Gaussian elimination method to solve the m-systems of 2^n equations in m unknowns has less complexity (3) than the complexity (2) of solving the same systems of equations. Therefore we calculate and update in Sect. 4.1 the complexities of REA-equivalence types I and IV given in [3] by using Lemma 1. Before that, we give a handy result that is used in our procedures.

Lemma 2 *[3]. Any linear function $L : \mathbb{F}_2^n \to \mathbb{F}_2^m$ can be converted to a matrix with n evaluations of L.*

4   REA-Equivalence Types

In this section, the verification procedures of some REA-equivalence types are presented with their complexities. We note that the required number of field operations for the multiplication of the matrix $M \in \mathbb{F}_2^{m \times n}$ with vector $V \in \mathbb{F}_2^{n \times 1}$ given by $M \cdot V$ is

$$mn - m \text{ XORs}, \ mn \text{ ANDs}. \tag{4}$$

The required number of field operations for the addition of the vectors $V_1, V_2 \in \mathbb{F}_2^{m \times 1}$ given by $V_1 + V_2$ is

$$m \text{ XORs}. \tag{5}$$

We use the standard multiplication algorithm for computing the matrix-vector multiplication. Note that there are various multiplication algorithms for the matrix-vector multiplication. One can use one of them to obtain a better complexity.

4.1   Verification of REA-Equivalence Types

In this part we revise the verification procedures of REA-equivalence types I to IV given in [3]. We show that finding the matrices for REA-equivalence is not enough and it also requires to check whether they satisfy the corresponding REA-equivalence (see Remark 2). We update their complexities, and particularly we apply Lemma 1 for types I and IV.

The complexities of REA-equivalence types are calculated in terms of the number of polynomial evaluations, field additions modulo 2 (*XORs*) and field multiplications modulo 2 (*ANDs*). Notice that the complexity of verification process for small n, m is significant since they are used in S-boxes. Thus, we count explicitly the number of field operations in the verification process and present the explicit complexity in case small n and m. In Sect. 4.3 we also present the complexity for large values of n, m in the big-O notation.

Notice that every vectorial Boolean function $F : \mathbb{F}_2^n \to \mathbb{F}_2^m$ can be written in the form

$$F(x) = F'(x) + F(0) \tag{6}$$

where $F'(x)$ has terms of algebraic degree at least 1. We define $k_{F,v}$ for any given function $F : \mathbb{F}_2^n \to \mathbb{F}_2^m$ and any given set $V \subseteq \text{Im}(F)$ as

$$k_{F,v} := |\{y \in \mathbb{F}_2^n | F(y) = v\}|, \tag{7}$$

for $v \in V$. The integer 2^j represents a vector of length m over $\mathbb{F}_2$ having 1 in the $(j+1)$-th component and 0 elsewhere for $j < m$ and $j, m \in \mathbb{N}$. We define the set $\mathcal{K}_1 = \{2^j | 0 \le j \le m - 1\} \subseteq \mathbb{F}_2^m$ for given $m \in \mathbb{Z}^+$.

Proposition 1. *Let* $F, G : \mathbb{F}_2^n \to \mathbb{F}_2^m$ *be vectorial Boolean functions defined by (6). Then the complexity of checking* F *and* G *for REA-equivalence of type I*

$$F(x) = M_1 \cdot G(x) + V_1 \text{ is} \tag{8}$$

i. $2\,km^2$ *XORs,* $2\,km^2$ *ANDs and* $2^{n+1} + 2$ *evaluations in case* $\mathcal{K}_1 \subseteq \text{Im}(G')$ *where* $k = \prod\limits_{v \in \mathcal{K}_1} k_{G',v}$.

ii. $(\dfrac{m^2(2k+1) + m}{2})(2^n - 1) + 2\,km^2$ *XORs,* $2\,km^2$ *ANDs and* $2^{n+1} + 2$ *evaluations in case* G *is arbitrary where*

$$k = \prod\limits_{v \in \text{Im}(G')} k_{G',v}.$$

Proof. Using (6), the REA-equivalence of type I can be written in the following form

$$F'(x) + F(0) = M_1 \cdot G'(x) + M_1 \cdot G(0) + V_1,$$

and it gives us the following equations;

$$\begin{cases} F(0) &= M_1 \cdot G(0) + V_1, \\ F'(x) &= M_1 \cdot G'(x). \end{cases} \tag{9}$$

Let us find m by m matrix M_1 from the second equation in (9). There exist two different cases for the function G'. If the set $\mathcal{K}_1 \subseteq \mathrm{Im}(G')$, then there exists at least one $x \in \mathbb{F}_2^n$ such that $G'(x) = 2^j$ for all $j \in \{0, \ldots, m-1\}$. Notice that if not, we should interchange F and G. To find those values of $x \in \mathbb{F}_2^n$, we need to compute the values $G'(x)$ for all $x \in \mathbb{F}_2^n$ with 2^n evaluations of G'. Next, we compute the values $F'(x)$ for those $x \in \mathbb{F}_2^n$ to find matrix M_1 with at most 2^n evaluations of F'. That is, the $j+1$-th column of M_1 can be found by means of the values $F'(x) = M_1 \cdot 2^j = M_{1_{j+1}}$ such that $G'(x) = 2^j$ for all $j \in \{0, \ldots, m-1\}$ by Lemma 2:

$$F'(x) = M_{1_{j+1}} = \begin{pmatrix} m_{1,j+1} \\ m_{2,j+1} \\ \vdots \\ m_{m,j+1} \end{pmatrix}.$$

If there are more than one x satisfying $G'(x) = 2^j$ for $j \in \{0, \ldots, m-1\}$, then there may be more than one $j+1$-th column of matrix M_1. On the basis of this, there may be more than one the candidate matrices M_1. Notice that the number of these matrices M_1 is k, which is the product of the size of the inverse image of the elements 2^j in the set $\mathrm{Im}(G')$ for all $j \in \{0, \ldots, m-1\}$. Therefore, the complexity of finding M_1 is 2^{n+1} evaluations.

For a given matrix M_1, one can easily compute the vector V_1 from the first equation in (9) as follows. One needs 2 evaluations of $F(0)$ and $G(0)$. The required number of field operations of the matrix-vector product $M_{1_{m \times m}} \cdot G(0)_{m \times 1}$ as in (4) is $m^2 - m$ XORs, m^2 ANDs. The required number of field addition for $V_1 = M_1 \cdot G(0) + F(0)$ is m XORs. Therefore, the complexity of finding V_1 is m^2 XORs, m^2 ANDs and 2 evaluations.

Moreover, we need to verify (see Remark 2) whether the given matrices M_1 and V_1 satisfy the Eq. (8). The complexity of checking process is m^2 XORs and m^2 ANDs. For each matrix M_1, we need to repeat these processes. The number of this repetition is at most k. If the Eq. (8) is satisfied for some M_1 and V_1, two functions F and G are REA-equivalent of type I; otherwise, they are REA-inequivalent of type I. As a result, the total complexity of verification for REA-equivalence of F and G is $2\,km^2$ XORs, $2\,km^2$ ANDs and $2^{n+1} + 2$ evaluations. This completes the proof of part (i).

Proof of part (ii): Let now G be arbitrary and $F'(x)_i$ be the i-th component of $F'(x)$. Denote $\mathrm{Im}(G')$ the image set of G' and $u_{G'} = |\mathrm{Im}(G')|$. Let also $N_{G'}$ be any subset of $\mathbb{F}_2^n$ which satisfies $|N_{G'}| = u_{G'}$ and $|\{G'(a)|a \in N_{G'}\}| = u_{G'}$. To begin with, we need to compute the values of $F'(x)$ and $G'(x)$ for all $x \in \mathbb{F}_2^n$ with 2^{n+1} evaluations. To find the matrix M_1, we construct the system of equations separately for each row of M_1. We use $u_{G'}$ many $x \in N_{G'}$ out of 2^n to construct the system of equations. To find i-th row of M_1 for all $i \in \{1, \ldots, m\}$, the system of equations can be constructed as follows for all $x_j \in N_{G'}, 1 \le j \le u_{G'}$:

$$F'(x_1)_i = m_{i,1}G'(x_1)_1 \oplus m_{i,2}G'(x_1)_2 \oplus \cdots \oplus m_{i,m}G'(x_1)_m$$
$$F'(x_2)_i = m_{i,1}G'(x_2)_1 \oplus m_{i,2}G'(x_2)_2 \oplus \cdots \oplus m_{i,m}G'(x_2)_m$$
$$\vdots \hspace{8cm} (10)$$
$$F'(x_{u_{G'}})_i = m_{i,1}G'(x_{u_{G'}})_1 \oplus m_{i,2}G'(x_{u_{G'}})_2 \oplus \cdots \oplus m_{i,m}G'(x_{u_{G'}})_m$$

The matrix representation of this system is

$$\begin{pmatrix} F'(x_1)_i \\ F'(x_2)_i \\ \vdots \\ F'(x_{u_{G'}})_i \end{pmatrix} = \begin{pmatrix} G'(x_1)_1 & G'(x_1)_2 & \cdots & G'(x_1)_m \\ G'(x_2)_1 & G'(x_2)_2 & \cdots & G'(x_2)_m \\ \vdots & \vdots & \ddots & \vdots \\ G'(x_{u_{G'}})_1 & G'(x_{u_{G'}})_2 & \cdots & G'(x_{u_{G'}})_m \end{pmatrix} \cdot \begin{pmatrix} m_{i,1} \\ m_{i,2} \\ \vdots \\ m_{i,m} \end{pmatrix} \qquad (11)$$

equivalently, $F_i' = \mathbf{G}' \cdot M_{1_i}$ where $\mathbf{G}'$ is the coefficients matrix of the system for all $i \in \{1, \ldots, m\}$. The augmented matrix of the systems (11) for i-th row of M_1 is $[\mathbf{G}'|F_i']$. The systems (11) for all $i \in \{1, \ldots, m\}$ can be solved in a unique augmented matrix because of the same $\mathbf{G}'$. The augmented matrix of the systems (11) for all rows of M_1 is

$$[\mathbf{G}'|N_{G'}] = [\mathbf{G}'|F_1'|F_2'|\cdots|F_m']. \qquad (12)$$

On the other hand, if G' is not one-to-one function, it requires to find all sets $N_{G'}$ satisfying $|N_{G'}| = u_{G'}$ and $|\{G'(a)|a \in N_{G'}\}| = u_{G'}$ so that we can find all the candidate matrices M_1. Because for distinct sets $N_{G'}$ there may be distinct matrices M_1. The number of sets $N_{G'}$ is k which is the product of the size of the inverse image of all elements in the set $\mathrm{Im}(G')$. Accordingly, there exist k distinct augmented matrices

$$[\mathbf{G}'|N_{G'}^l] \qquad (13)$$

which have the same coefficient matrix $\mathbf{G}'$ as in (12) for all $l \in \{1, \ldots, k\}$. Notice that, there may be distinct matrices M_1 corresponding to each set $N_{G'}^l$ if the system (13) is consistent for each l. Thus, the number of the candidate matrices M_1 is less than or equals to k. The systems (13) for all $l \in \{1, \ldots, k\}$ can be solved in a unique augmented matrix. The augmented matrix of the k distinct systems for all $N_{G'}$ sets is

$$[\mathbf{G}'|N_{G'}^1|N_{G'}^2|\cdots|N_{G'}^k] \qquad (14)$$

and this can be solved by simultaneous application of the Gaussian elimination method with $(\frac{m^2(2k+1)+m}{2})(2^n - 1)$ XORs by Lemma 1. This cost and 2^{n+1} evaluations give the cost of finding the candidate matrices M_1. Similar to part (i), the total complexity of computing V_1 and checking process for all the candidate matrices M_1 is $2\,km^2$ XORs $2\,km^2$ ANDs and 2 evaluations. Therefore, the total complexity of verification for REA-equivalence of F and G is $(\frac{m^2(2k+1)+m}{2})$ $(2^n - 1) + 2\,km^2$ XORs, $2\,km^2$ ANDs and $2^{n+1} + 2$ evaluations.   □

Corollary 1. *Let $n = m$ and G be a permutation. Then the complexity of checking F and G for REA-equivalence of type I is $2n^2$ XORs, $2n^2$ ANDs and $2^n + n + 2$ evaluations.*

Proof. Since G is a permutation, the function G' satisfies part (i) of Proposition 1, and $k = 1$. Hence, the proof follows from part (i) of Propositions 1. $\square$

Remark 1. The reason why we use only the elements of the set $N_{G'}$ in $\mathbb{F}_2^n$ to construct the system (10) is the following. If we use $x_0 \in N_{G'}$ and $x_1 \notin N_{G'}$ such that $G'(x_0) = G'(x_1)$, there exist two cases: the system is inconsistent in case $F'(x_0) \neq F'(x_1)$, and some row operations would be unnecessary in case $F'(x_0) = F'(x_1)$.

Remark 2. We urge to explain why the checking process is necessary after finding the matrices in the verification procedure of REA-equivalence types. In particular, we consider the REA-equivalence Eq. (8) given in Proposition 1. To verify whether F and G are REA-equivalent or not, we solve (8) for given F and G. Namely, we try to determine the invertible matrix M_1 and the vector V_1 satisfying (8). For this purpose, (8) is divided into sub-equations as given in (9). To determine M_1 and V_1 from (9), we evaluate F', G' at selected values of $x \in \mathbb{F}_2^n$ and evaluate F, G at $x = 0$, respectively. So, M_1 and V_1 have been found. Since some of x values do not contribute to finding these matrices, the REA-equivalence may not hold at these values. In such a case, F and G are REA-inequivalent functions for the matrices M_1 and V_1 for type I.

We now present an example showing that only solving the system of equations for type I is not enough to decide whether two functions are equivalent or not.

Example 1. Let $F, G : \mathbb{F}_2^3 \longrightarrow \mathbb{F}_2^3$ be vectorial Boolean functions and $F(x) = (x_3, x_1 \oplus x_2 \oplus 1, x_1 \oplus 1)$ and $G(x) = (x_1 x_3 \oplus 1, x_1, x_2)$, where $x = (x_3, x_2, x_1) \in \mathbb{F}_2^3$. We will show that functions F and G are REA-inequivalent of type I. First of all, REA-equivalence of type I can be written as in (9) where $F'(x) = (x_3, x_1 \oplus x_2, x_1)$ and $G'(x) = (x_1 x_3, x_1, x_2)$, $F(0) = (0, 1, 1)$ and $G(0) = (1, 0, 0)$. Next, we check whether part (i) is satisfied or not. It is obvious that G' does not satisfy part (i) since $2^0 \notin img(G')$. On the other hand, F' satisfies part(i) but the solution

$$M_1 = \begin{pmatrix} 0 & 0 & 0 \\ 0 & 0 & 1 \\ 0 & 1 & 1 \end{pmatrix}$$

is not invertible matrix over $\mathbb{F}_2$. So, F is REA-inequivalent to G of type I.

We also verify the inequality by using part (ii). We first constitute the sets $N_{G'} \subseteq \mathbb{F}_2^3$ by taking into account the elements of $Im(G')$. Then, the system of equations can be constructed by using four distinct sets $N_{G'}$. The augmented matrix of the systems is $[\mathbf{G'}|N_{G'}^1|N_{G'}^2|N_{G'}^3|N_{G'}^4]$:

$$\mathbf{Aug} = \begin{pmatrix} 000 & 000 & 000 & 100 & 100 \\ 010 & 011 & 011 & 011 & 011 \\ 001 & 010 & 110 & 010 & 110 \\ 011 & 001 & 001 & 001 & 001 \\ 110 & 111 & 111 & 111 & 111 \\ 111 & 101 & 101 & 101 & 101 \end{pmatrix} \sim \begin{pmatrix} 100 & 100 & 100 & 100 & 100 \\ 010 & 011 & 011 & 011 & 011 \\ 001 & 010 & 110 & 010 & 110 \\ 000 & 000 & 100 & 000 & 100 \\ 000 & 000 & 000 & 100 & 100 \\ 000 & 000 & 000 & 000 & 000 \end{pmatrix}.$$

$$(15)$$

It can be reduced to row reduced echelon form by simultaneous Gaussian elimination method as above. There exists a unique solution

$$M_1 = \begin{pmatrix} 1 & 0 & 0 \\ 0 & 1 & 1 \\ 0 & 1 & 0 \end{pmatrix}.$$

Then we compute $V_1 = F(0) + M_1 \cdot G(0) = (1,1,1)$. Finally, we need to check whether the Eq. (8) for given M_1, V_1 holds. We define $K(x) := M_1 \cdot G(x) + V_1$ and substitute the values M_1, V_1 into $K(x)$,

$$K(x) := \begin{pmatrix} 1 & 0 & 0 \\ 0 & 1 & 1 \\ 0 & 1 & 0 \end{pmatrix} \cdot \begin{pmatrix} x_1 x_3 \oplus 1 \\ x_1 \\ x_2 \end{pmatrix} + \begin{pmatrix} 1 \\ 1 \\ 1 \end{pmatrix} = (x_1 x_3, x_1 \oplus x_2 \oplus 1, x_1 \oplus 1).$$

Since $K(x) \neq F(x)$, we say that F is REA-inequivalent to G of type I.

Example 2. Let $F, G : \mathbb{F}_2^3 \longrightarrow \mathbb{F}_2^3$ be vectorial Boolean functions and $F(x) = (x_1 \oplus x_2 \oplus x_3 \oplus x_1 x_2 \oplus 1, x_2 \oplus x_1 x_3 \oplus x_2 x_3, x_1 \oplus x_2 x_3 \oplus 1)$ and $G(x) = (x_1 \oplus x_2 \oplus x_3 \oplus x_1 x_2, x_1 \oplus x_2 x_3 \oplus 1, x_1 \oplus x_2 \oplus x_1 x_3)$ where $x = (x_3, x_2, x_1) \in \mathbb{F}_2^3$. Functions F and G are REA-equivalent of type I. Notice that *the Univariate polynomial representation* of these functions are $F(x) = x^3 + \alpha^2 + 1$ and $G(x) = x^6 + \alpha$ where α is a root of primitive polynomial $f(x) = x^3 + x + 1$ over $\mathbb{F}_2$. When these functions are applied to REA-equivalence of types I, IV, V, VI in MAGMA [2], it is seen that they are REA-equivalent. We note that MAGMA codes can be found in [9].

Proposition 2. *Let $F, G : \mathbb{F}_2^n \to \mathbb{F}_2^n$ be vectorial Boolean functions and G be a permutation. Then the complexity of checking F and G for REA-equivalence of type II, $F(x) = G(M_2 \cdot x + V_2)$, is n^2 XORs, n^2 ANDs and $n+1$ evaluations.*

Proof. The equation $F(x) = G(M_2 \cdot x + V_2)$ can be written as $G^{-1}(F(x)) = M_2 \cdot x + V_2$ since G is a permutation. We define $H(x) := G^{-1}(F(x))$. Using (6), write $H(x) = H'(x) + H(0) = M_2 \cdot x + V_2$ and this gives

$$\begin{cases} H'(x) = M_2 \cdot x, \\ H(0) = V_2. \end{cases} \tag{16}$$

The vector V_2 can be found with 1 evaluation. For $x = 2^j$ find n by n matrix M_2 from (16) by $H'(2^j) = M_2 \cdot 2^j = M_{2_{j+1}}$ for all $j \in \{0, 1, \ldots, n-1\}$ with n evaluations as in Lemma 2. In addition, the equation

$$H(x) = M_2 \cdot x + V_2 \tag{17}$$

must be checked for M_2 and V_2, and this costs n^2 XORs, n^2 ANDs. Consequently, the total complexity of verification for REA-equivalence of F and G is n^2 XORs, n^2 ANDs and $n+1$ evaluations. $\square$

Example 3. Let $F, G : \mathbb{F}_2^3 \longrightarrow \mathbb{F}_2^3$ be vectorial Boolean functions and $F(x) = (x_1 x_2 \oplus x_1, x_3, x_2)$ and $G(x) = (x_1, x_3, x_2)$, where $x = (x_3, x_2, x_1) \in \mathbb{F}_2^3$. Then $H(x) = (x_3, x_2, x_1 x_2 \oplus x_1)$. There exist the solutions, M_2 and V_2, of the Eq. (16), but they do not satisfy the Eq. (17) of type II. Thus, we say that they are REA-inequivalent of type II.

Proposition 3. *Let $F, G : \mathbb{F}_2^n \to \mathbb{F}_2^m$ be vectorial Boolean functions. Then the complexity of checking F and G for REA-equivalence of type III, $F(x) = G(x) + M_3 \cdot x + V_1$, is mn XORs, mn ANDs and $n + 1$ evaluations.*

Proof. Denote $H(x) = F(x) + G(x)$. Then the REA-equivalence of type III, takes the form $H(x) = M_3 \cdot x + V_1$. Using (6), this can be rewritten as follows;

$$\begin{cases} H'(x) = M_3 \cdot x, \\ H(0) = V_1. \end{cases} \tag{18}$$

The matrices M_3 and V_1 can be found as in Proposition 2 with n evaluations and 1 evaluation, respectively. In addition, the equation

$$H(x) = M_3 \cdot x + V_1 \tag{19}$$

must be checked for a given M_3 and V_1 with mn XORs and mn ANDs. Consequently, the total complexity of verification for the equivalence of F and G is mn XORs, mn ANDs and $n + 1$ evaluations. $\qquad\square$

Example 4. Let $F, G : \mathbb{F}_2^3 \longrightarrow \mathbb{F}_2^3$ be vectorial Boolean functions and $F(x) = (x_3 \oplus 1, x_1 x_2, x_1)$ and $G(x) = (x_2 \oplus x_3, x_2, x_1 x_3)$, where $x = (x_3, x_2, x_1) \in \mathbb{F}_2^3$. Then $H(x) = (x_2 \oplus 1, x_1 x_2 \oplus x_2, x_1 \oplus x_1 x_3)$. There exist the solutions, M_3 and V_1, of the Eq. (18), but they do not satisfy the Eq. (19) of type III. Thus, we say that they are REA-inequivalent of type III.

Notice that every vectorial Boolean function $F : \mathbb{F}_2^n \to \mathbb{F}_2^m$ can be written in the form

$$F(x) = F'(x) + L_F(x) + F(0), \tag{20}$$

where F' has terms of algebraic degree at least 2 and L_F is the linear term.

Proposition 4. *Let $F, G : \mathbb{F}_2^n \to \mathbb{F}_2^m$ be vectorial Boolean functions and G be defined by (20). The complexity of checking F and G for REA-equivalence of type IV, $F(x) = M_1 \cdot G(x) + M_3 \cdot x + V_1$, is*

i. $k(nm^2 + 2m^2 + mn)$ XORs, $k(nm^2 + 2m^2 + mn)$ ANDs and $2^{n+1} + 2n + 2$ evaluations in case the set $\mathcal{K}_1 \subseteq \mathrm{Im}(G')$ where

$$k = \prod_{v \in \mathcal{K}_1} k_{G', v}.$$

ii. $(\dfrac{m^2(2k+1) + m}{2})(2^n - 1) + 2km^2$ XORs, $2km^2$ ANDs and $2^{n+1} + 2$ evaluations in case G is arbitrary where

$$k = \prod_{v \in \mathrm{Im}(G')} k_{G', v}.$$

Proof. Using (20), the REA-equivalence of type IV can be rewritten as follows

$$F'(x) + L_F(x) + F(0) = M_1 \cdot G'(x) + M_1 \cdot L_G(x) + M_3 \cdot x + M_1 \cdot G(0) + V_1 \quad (21)$$

and the following system of equations can be obtained from (21)

$$\begin{cases} F'(x) = M_1 \cdot G'(x), \\ L_F(x) = M_1 \cdot L_G(x) + M_3 \cdot x, \\ F(0) = M_1 \cdot G(0) + V_1. \end{cases} \quad (22)$$

To begin with, we find the matrix M_1 from the first equation in (22). This equation leads to two different cases for the function G' as in Proposition 1. First, we prove part (i). The method and the complexity of finding m by m matrix M_1 are the same as part (i) in Proposition 1. Thus, the complexity of finding the candidate matrices M_1 is 2^{n+1} evaluations. Notice that the number of the candidate matrices M_1 is k. For each given matrix M_1, the matrix M_3 and the vector V_1 can be found from (22). For $x = 2^j$ find M_3 by

$$L_F(2^j) + M_1 \cdot L_G(2^j) = M_3 \cdot 2^j = M_{3_{j+1}}$$

for all $j \in \{0, 1, \ldots, n-1\}$ with complexity nm^2 XORs, nm^2 ANDs and $2n$ evaluations. Moreover, $V_1 = F(0) + M_1 \cdot G(0)$ can be computed with complexity m^2 XORs, m^2 ANDs and 2 evaluations. Finally, we need to verify whether the matrices M_1, M_3 and V_1 satisfy the equation

$$F(x) = M_1 \cdot G(x) + M_3 \cdot x + V_1. \quad (23)$$

This costs $m^2 + mn$ XORs and $m^2 + mn$ ANDs. For each M_1, we need to repeat the process of finding V_1, M_3 and checking the Eq. (23) to verify whether they are REA-equivalent. The number of this repetition is at most k. If the Eq. (23) is satisfied for some M_1, M_3 and V_1, then we say that the functions F and G are REA-equivalent of type IV; otherwise, they are REA-inequivalent of type IV. Consequently, the total complexity of verification for the equivalence of F and G is $k(nm^2 + 2m^2 + mn)$ XORs, $k(nm^2 + 2m^2 + mn)$ ANDs and $2^{n+1} + 2n + 2$ evaluations.

Proof of part (ii): The method and the complexity of finding M_1 in case G arbitrary are the same as part (ii) in Proposition 1. For each M_1, we need to repeat the process of finding V_1, M_3 and checking the Eq. (23) as in part(i). Consequently, the total complexity of verification for the equivalence of F and G is $(\frac{m^2(2k+1)+m}{2})(2^n - 1) + k(nm^2 + 2m^2 + mn)$ XORs, $k(nm^2 + 2m^2 + mn)$ ANDs and $2^{n+1} + 2n + 2$ evaluations.   □

Example 5. Let $F, G : \mathbb{F}_2^3 \longrightarrow \mathbb{F}_2^3$ be vectorial Boolean functions and $F(x) = (x_3 \oplus x_2x_3 \oplus 1, x_1 \oplus x_2 \oplus x_1x_2 \oplus 1, x_1x_3 \oplus x_1x_2x_3)$ and $G(x) = (x_1 \oplus x_3 \oplus x_1x_2, x_2 \oplus x_1x_3 \oplus x_2x_3 \oplus 1, x_1x_2x_3)$, where $x = (x_3, x_2, x_1) \in \mathbb{F}_2^3$. There exist the solutions, M_1, M_3 and V_1, of the Eq. (22), but they do not satisfy the Eq. (23) of type IV. Thus, we say that they are REA-inequivalent of type IV.

The summary of updated complexities of REA-equivalence types introduced in [3] is presented in Table 2.

Table 2. Updated complexities of REA-equivalence types

Type	REA-equivalence	Complexity in case G Arbitrary		
		#XOR	#AND	#Evaluation
I	$F(x) = M_1 \cdot G(x) + V_1$	$(2km^2 + m^2 + m)(2^{n-1} - \frac{1}{2})$ $+ 2km^2$	$2km^2$	$2^{n+1} + 2$
III	$F(x) = G(x) + M_3 \cdot x + V_1$	mn	mn	$n+1$
IV	$F(x) = M_1 \cdot G(x) + M_3 \cdot x + V_1$	$(2km^2 + m^2 + m)(2^{n-1} - \frac{1}{2})$ $+ k(nm^2 + 2m^2 + mn)$	$k(nm^2 + 2m^2 + mn)$	$2^{n+1} + 2n + 2$
		Complexity in case $\mathcal{K}_1 = \{2^j \mid 0 \le j \le m-1\} \subseteq \mathrm{Im}(G')$		
I	$F(x) = M_1 \cdot G(x) + V_1$	$2km^2$	$2km^2$	$2^{n+1} + 2$
IV	$F(x) = M_1 \cdot G(x) + M_3 \cdot x + V_1$	$k(nm^2 + 2m^2 + mn)$	$k(nm^2 + 2m^2 + mn)$	$2^{n+1} + 2n + 2$
		Complexity in case G Permutation		
I	$F(x) = M_1 \cdot G(x) + V_1$	$2n^2$	$2n^2$	$2^n + n + 2$
II	$F(x) = G(M_2 \cdot x + V_2)$	n^2	n^2	$n+1$

4.2 Verification of New Types of REA-equivalence

We introduce two new types of REA-equivalence in Table 3. The verification procedures of these types, namely types V and VI, with their complexities are constructed in this section.

Table 3. New types of REA-equivalence

Types	REA-equivalence
V	$F(x) = G(M_2 \cdot x) + M_3 \cdot x + V_1$
VI	$F(x) = G(M_2 \cdot x) + V_1$

Proposition 5. *Let $F, G : \mathbb{F}_2^n \to \mathbb{F}_2^m$ be vectorial Boolean functions defined by (20). Then the complexity of checking F and G for REA-equivalence of type V, $F(x) = G(M_2 \cdot x) + M_3 \cdot x + V_1$ such that M_2 is a permutation matrix, is $k(n^2 + 2mn + m - n) + m$ XORs, $k(n^2 + mn)$ ANDs and $2^n + 2n + 2 + k(n+1)$ evaluations in case $\mathcal{K}_2 = \{F'(2^j) \mid 0 \le j \le n-1\} \subseteq \mathrm{Im}(G')$ where*

$$k = \prod_{v \in \mathcal{K}_2} k_{G', v}.$$

Proof. Using (20), the REA-equivalence of type V can be rewritten as follows

$$F'(x) + L_F(x) + F(0) = G'(M_2 \cdot x) + L_G(M_2 \cdot x) + M_3 \cdot x + G(0) + V_1. \quad (24)$$

Since M_2 is a permutation matrix, the following system is obtained from (24):

$$\begin{cases} F'(x) = G'(M_2 \cdot x) \\ L_F(x) = L_G(M_2 \cdot x) + M_3 \cdot x \\ F(0) = G(0) + V_1. \end{cases} \quad (25)$$

One can easily compute $V_1 = F(0) + G(0)$ with m *XORs* and 2 evaluations. Let us find M_2 from the first equation in (25). It is clear that the set $\mathcal{U} = \{2^j | 0 \leq j \leq n - 1\}$ is a subset of $\mathbb{F}_2^n$. To find the candidate $j + 1$-th column of M_2, $M_2 \cdot 2^j = M_{2_{j+1}}$, we should solve the equation

$$G'(M_{2_{j+1}}) = F'(2^j) \tag{26}$$

for all $2^j \in \mathcal{U}$. To solve the Eq. (26), we need to compute the values $F'(2^j)$ for all $2^j \in \mathcal{U}$ and $G'(y)$ for all $y \in \mathbb{F}_2^n$ with $2^n + n$ evaluations. Thus, the candidate $j + 1$-th columns of M_2 are found by means of $y = M_{2_{j+1}}$ which satisfies the Eq. (26) for $j \in \{0, 1, \ldots, n - 1\}$. Notice that the number of the candidate matrices M_2 is k, which is the product of the number of y. For each given matrix M_2, one can easily compute M_3 from the second equation in (25) by using $L_F(2^j) + L_G(M_{2_{j+1}}) = M_{3_{j+1}}$ for all $2^j \in \mathcal{U}$. The complexity of finding the matrix M_3 is mn *XORs* and $2n$ evaluations. Next, we need to verify whether the matrices M_2, M_3 and V_1 satisfy the equation

$$F(x) = G(M_2 \cdot x) + M_3 \cdot x + V_1. \tag{27}$$

This costs $n^2 + mn + m - n$ *XORs*, $n^2 + mn$ *ANDs* and 1 evaluation. For each given matrix M_2, we need to repeat the process of finding M_3 and checking the Eq. (27) to verify whether they are REA-equivalent. The number of this repetition is at most k. If the Eq. (27) is satisfied for some M_2, M_3 and V_1, the functions F and G are REA-equivalent of type V; otherwise, they are REA-inequivalent of type V. Consequently, the total complexity of this verification is $k(n^2 + 2mn + m - n) + m$ *XORs*, $k(n^2 + mn)$ *ANDs* and $2^n + 2n + 2 + k(n + 1)$ evaluations. $\qquad\square$

Example 6. Let $F, G : \mathbb{F}_2^3 \longrightarrow \mathbb{F}_2^3$ be vectorial Boolean functions and $F(x) = (x_1 \oplus x_2 \oplus x_3 \oplus x_1 x_2 \oplus 1, x_2 \oplus x_1 x_3, x_1 \oplus x_2 x_3 \oplus 1)$ and $G(x) = (x_1 \oplus x_2 \oplus x_3 \oplus x_1 x_2, x_1 \oplus x_2 x_3 \oplus 1, x_1 \oplus x_2 \oplus x_1 x_3)$, where $x = (x_3, x_2, x_1) \in \mathbb{F}_2^3$. We will show that functions F and G are REA-equivalent of type V. First of all, REA-equivalence of type V can be rewritten as in (25) where $F'(x) = (x_1 x_2, x_1 x_3, x_2 x_3)$, $G'(x) = (x_1 x_2, x_2 x_3, x_1 x_3)$, $L_F(x) = (x_1 \oplus x_2 \oplus x_3, x_2, x_1)$, $L_G(x) = (x_1 \oplus x_2 \oplus x_3, x_1, x_1 \oplus x_2)$, $F(0) = (1, 0, 1)$ and $G(0) = (0, 1, 0)$.

It is obvious that $V_1 = (1, 1, 1)$. To find the candidate $j + 1$-th column $M_{2_{j+1}}$ of the matrix M_2, we first compute the values $F'(2^j)$ and find $y \in \mathbb{F}_2^3$ satisfying $G'(y) = F(2^j)$ for $j = 0, 1, 2$. The candidate j-th columns of M_2 are the same values $(0, 0, 0), (0, 0, 1), (0, 1, 0), (1, 0, 0)$ for all $j = 0, 1, 2$. Hence, there exist 64 distinct candidate matrices M_2; however only six of them are the permutation matrices:

$$\begin{pmatrix} 0 & 0 & 1 \\ 0 & 1 & 0 \\ 1 & 0 & 0 \end{pmatrix}, \begin{pmatrix} 0 & 1 & 0 \\ 0 & 0 & 1 \\ 1 & 0 & 0 \end{pmatrix}, \begin{pmatrix} 0 & 0 & 1 \\ 1 & 0 & 0 \\ 0 & 1 & 0 \end{pmatrix}, \begin{pmatrix} 0 & 1 & 0 \\ 1 & 0 & 0 \\ 0 & 0 & 1 \end{pmatrix}, \begin{pmatrix} 1 & 0 & 0 \\ 0 & 1 & 0 \\ 0 & 0 & 1 \end{pmatrix} \text{ and } \begin{pmatrix} 1 & 0 & 0 \\ 0 & 0 & 1 \\ 0 & 1 & 0 \end{pmatrix}.$$

The first candidate matrix

$$M_2 = \begin{pmatrix} 0\,0\,1 \\ 0\,1\,0 \\ 1\,0\,0 \end{pmatrix} \text{ gives } M_3 = \begin{pmatrix} 0\,0\,0 \\ 1\,1\,0 \\ 1\,1\,1 \end{pmatrix}.$$

Then, we need to check whether the Eq. (27) for given M_2, M_3, V_1 holds. We define $K(x) := G(M_2 \cdot x) + M_3 \cdot x + V_1$ and substitute the values M_2, M_3, V_1 into $K(x)$,

$$K(x) := G\left(\begin{pmatrix} 0\,0\,1 \\ 0\,1\,0 \\ 1\,0\,0 \end{pmatrix} \cdot \begin{pmatrix} x_3 \\ x_2 \\ x_1 \end{pmatrix} \right) + \begin{pmatrix} 0\,0\,0 \\ 1\,1\,0 \\ 1\,1\,1 \end{pmatrix} \cdot \begin{pmatrix} x_3 \\ x_2 \\ x_1 \end{pmatrix} + \begin{pmatrix} 1 \\ 1 \\ 1 \end{pmatrix}$$
$$= (x_1 \oplus x_2 \oplus x_3 \oplus x_2 x_3 \oplus 1, x_2 \oplus x_1 x_2, x_1 \oplus x_1 x_3 \oplus 1).$$

Since $K(x) \neq F(x)$, F is REA-inequivalent to G for this matrix M_2. However, we can not say that F is REA-inequivalent to G for type V since there exist more than one candidate matrices M_2. The next matrix

$$M_2 = \begin{pmatrix} 1\,0\,0 \\ 0\,0\,1 \\ 0\,1\,0 \end{pmatrix} \text{ gives } M_3 = \begin{pmatrix} 0\,0\,0 \\ 0\,0\,0 \\ 0\,1\,0 \end{pmatrix}.$$

It is easy to see that the Eq. (27) for these matrices M_2, M_3, V_1 holds. Therefore, the function F is REA-equivalent to G for type V.

Proposition 6. *Let $F, G : \mathbb{F}_2^n \to \mathbb{F}_2^m$ be vectorial Boolean functions defined by (6). Then the complexity of checking F and G for REA-equivalence of type VI, $F(x) = G(M_2 \cdot x) + V_1$, is $k(n^2 - n + m) + m$ XORs, kn^2 ANDs and $2^n + n + 2 + k$ evaluations in case $\mathcal{K}_2 = \{F'(2^j) | 0 \leq j \leq n - 1\} \subseteq \mathrm{Im}(G')$ where*

$$k = \prod_{v \in \mathcal{K}_2} k_{G', v}.$$

Proof. Using (6), the REA-equivalence of type VI can be rewritten as $F'(x) + F(0) = G'(M_2 \cdot x) + G(0) + V_1$ and it gives the following form

$$\begin{cases} F'(x) = G'(M_2 \cdot x) \\ F(0) = G(0) + V_1. \end{cases} \tag{28}$$

One can easily compute V_1 as in Proposition 2 with m XORs and 2 evaluations. The first equation in (28) can be solved by the same method as in Proposition 5 to find M_2. Thus, the candidate matrices M_2 can be found with $2^n + n$ evaluations. Notice that the number of the candidate matrices M_2 is k. For each given matrix M_2, we need to verify whether the matrices M_2 and V_1 satisfy the equation

$$F(x) = G(M_2 \cdot x) + V_1 \tag{29}$$

with complexity $k(n^2 - n + m)$ XORs, kn^2 ANDs and k evaluations. Consequently, the total complexity of verification for equivalence of F and G is $k(n^2 - n + m) + m$ XORs, kn^2 ANDs and $2^n + n + 2 + k$ evaluations. $\square$

Corollary 2. *Let $m = n$ and G be a permutation. Then the complexity of checking F and G for REA-equivalence of type VI is $n^2 + n$ XORs, n^2 ANDs and $2^n + n + 3$ evaluations.*

Proof. Since G is a permutation, $k = 1$. Hence, the proof follows from Proposition 6. □

Example 7. Let $F, G : \mathbb{F}_2^3 \longrightarrow \mathbb{F}_2^3$ be vectorial Boolean functions such that $F(x) = (x_1 \oplus x_2, x_3, x_2 \oplus 1)$ and $G(x) = (x_1, x_3 \oplus 1, x_2)$ where $x = (x_3, x_2, x_1) \in \mathbb{F}_2^3$. F and G are REA-equivalent of type VI.

We present the complexity of verification procedures of these REA-equivalence types in Table 4.

Table 4. The complexities of two new types of REA-equivalence

Type	REA-equivalence	Complexity in case $K_2 = \{F'(2^j) \mid 0 \leq j \leq n-1\} \subseteq \operatorname{Im}(G')$		
		#XOR	#AND	#Evaluation
V	$F(x) = G(M_2 \cdot x) + M_3 \cdot x + V_1$	$k(n^2 + 2mn + m - n) + m$	$k(n^2 + mn)$	$2^n + 2n + 2 + k(n+1)$
VI	$F(x) = G(M_2 \cdot x) + V_1$	$k(n^2 - n + m) + m$	kn^2	$2^n + n + 2 + k$
		Complexity in case G Permutation		
VI	$F(x) = G(M_2 \cdot x) + V_1$	$n^2 + n$	n^2	$2^n + n + 3$

4.3 Comparison with Previous Work

In [3] authors illustrate the complexity of verification procedures without checking whether the solution satisfies the equivalence relation or not, except for the part (i) of type I. However, as we explained in Remark 2, the checking process is necessary after finding the matrices in the verification procedure of any REA-equivalence types. Examples 1, 3, 4 and 5 illustrate that the determination of the matrices is itself does not fully complete the process and it is also necessary to check whether the found matrices satisfy the equation even if they are unique. Therefore we include the checking process into the verification procedures of REA-equivalence types. Secondly, Example 6 illustrates that checking REA-equivalence of two functions for only one $x \in \mathbb{F}_2^n$ which satisfies $G'(x) = F'(2^j)$ for some $j = 0, 1, 2$ may not complete the process. We need to use all such x to find all the candidate columns of matrix M_2. In this example, F and G are REA-inequivalent for some M_2 while they are REA-equivalent for some M_2. Thus, there may exist another matrix M for which F and G are REA-equivalent for types I, IV, V and VI while they are REA-inequivalent for one matrix M. The authors in [3] follow the verification procedures for types I and IV without taking into account that the matrix M_1 may be more than one. As a result, it is not completely fair to compare the present complexities with the ones in [3]. We present the complexities of REA-equivalence types given in [3] in Table 5.

In our computational complexity, we count the explicit number of field operations: *XOR*, *AND* and polynomial evaluation. The polynomial evaluation of vectorial Boolean functions can be done by *table look-up operation*, where the tables consist of values of the functions, and each table look-up operation can be considered as one field operation, and each XOR and AND operations can be considered as one field operation. These are meaningful as we are in $\mathbb{F}_2$. Thus, we can add the number of $XORs$, $ANDs$ and table look-up operations. This enables us to see the total number of operations of REA-equivalence procedures. For large n, m, the total complexities of REA-equivalence types are presented with big-O notation in Tables 6 and 7 so that the difference between our results and ones in [3] can be seen easily.

Table 5. The complexities of REA-equivalence types given in [3] where $\mathcal{K}_1 = \{2^j | 0 \leq j \leq m - 1\}$

Types	REA-equivalence	G Permutation	$\mathcal{K}_1 \subseteq \mathrm{Im}(G')$	G Arbitrary
I	$F(x) = M_1 \cdot G(x) + V_1$		$O(2^{n+1})$	$O(m2^{2n})$
II	$F(x) = G(M_2 \cdot x + V_2)$	$O(n)$		
III	$F(x) = G(x) + M_3 \cdot x + V_1$			$O(n)$
IV	$F(x) = M_1 \cdot G(x) + M_3 \cdot x + V_1$		$O(2^{n+1})$	$O(m2^{2n})$

Table 6. Updated complexities of REA-equivalence types for large n, m

Type	REA-equivalence	G Permutation	$\mathcal{K}_1 \subseteq \mathrm{Im}(G')$	G Arbitrary
I	$F(x) = M_1 \cdot G(x) + V_1$	$O(2^n)$	$O(2^n + km^2)$	$O(km^2 2^n)$
II	$F(x) = G(M_2 \cdot x + V_2)$	$O(n^2)$		
III	$F(x) = G(x) + M_3 \cdot x + V_1$			$O(mn)$
IV	$F(x) = M_1 \cdot G(x) + M_3 \cdot x + V_1$		$O(2^n + knm^2)$	$O(km^2 2^n)$

Table 7. The complexities of new REA-equivalence types for large n, m where $\mathcal{K}_2 = \{F'(2^j) | 0 \leq j \leq n - 1\}$

Type	REA-equivalence	G Permutation	$\mathcal{K}_2 \subseteq \mathrm{Im}(G')$
V	$F(x) = G(M_2 \cdot x) + M_3 \cdot x + V_1$		$O(2^n + k(n^2 + mn))$
VI	$F(x) = G(M_2 \cdot x) + V_1$	$O(2^n)$	$O(2^n + k(n^2 + m))$

5   Conclusion

This paper studies REA-equivalence algorithms. Firstly, we analyze the verification procedures of REA-equivalence types given in [3]. We update the complexities of the verification procedures and count explicitly the number of operations. Furthermore, we introduce two new types of REA-equivalence and construct their verification procedures with their complexities. As a result, this paper provides fast check procedures of vectorial Boolean functions for REA-equivalence. We also implemented our procedures in MAGMA for determining whether two functions are equivalent or not, and MAGMA codes can be found in [9].

Acknowledgment. We first thank the referees for providing detailed comments and suggestions. The second author is partially supported by the Scientific and Technological Research Council of Turkey (TÜBİTAK). The third author is supported by TÜBİTAK under the National Postdoctoral Research Scholarship No 2219.

References

1. Biryukov, A., De Canniere, C., Braeken, A., Preneel, B.: A tool-box for cryptanalysis: linear and affine equivalence algorithms. In: Biham, E. (ed.) Advances in Cryptology — EUROCRYPT 2003. LNCS, vol. 2656, pp. 33–50. Springer, Heidelberg (2003)
2. Bosma, W., Cannon, J., Playoust, C.: The Magma algebra system, I. The user language. J. Symb. Comput. **24**, 235–265 (1997)
3. Budaghyan, L., Kazymyrov, O.: Verification of restricted EA-equivalence for vectorial Boolean functions. In: Özbudak, F., Rodríguez-Henríquez, F. (eds.) WAIFI 2012. LNCS, vol. 7369, pp. 108–118. Springer, Heidelberg (2012)
4. Budaghyan, L., Carlet, C., Pott, A.: New classes of almost bent and almost perfect nonlinear polynomials. IEEE Trans. Inform. Theory **52**, 1141–1152 (2006)
5. Carlet, C., Charpin, P., Zinoviev, V.: Codes, bent functions and permutations suitable for DES-like cryptosystems. Des. Codes Crypt. **15**(2), 125–156 (1998)
6. Carlet, C.: Vectorial Boolean functions for cryptography. Boolean Model. Methods Math. Comput. Sci. Eng. **134**, 398–469 (2010)
7. Chabaud, F., Vaudenay, S.: Links between differential and linear cryptanalysis. In: De Santis, A. (ed.) EUROCRYPT 1994. LNCS, vol. 950, pp. 356–365. Springer, Heidelberg (1995)
8. Nyberg, K.: Differentially uniform mappings for cryptography. In: Helleseth, T. (ed.) EUROCRYPT 1993. LNCS, vol. 765, pp. 55–64. Springer, Heidelberg (1994)
9. Sınak, A.: On verification of restricted extended affine equivalence of vectorial Boolean functions. Master's thesis, Middle East Technical University (2012)
10. Williams, V.V.: Breaking the Coppersmith-Winograd barrier, November 2011

On o-Equivalence of Niho Bent Functions

Lilya Budaghyan$^{1(\boxtimes)}$, Claude Carlet^2, Tor Helleseth^1,
and Alexander Kholosha^1

^{1} Department of Informatics, University of Bergen, PB 7803, 5020 Bergen, Norway
{Lilya.Budaghyan,Tor.Helleseth,Alexander.Kholosha}@ii.uib.no
^{2} LAGA, UMR 7539, CNRS, Department of Mathematics, University of Paris 8
and University of Paris 13, 2 rue de la Liberté, 93526 Saint-Denis Cedex, France
claude.carlet@univ-paris8.fr

Abstract. As observed recently by the second author and S. Mesnager, the projective equivalence of o-polynomials defines, for Niho bent functions, an equivalence relation called o-equivalence. These authors also observe that, in general, the two o-equivalent Niho bent functions defined from an o-polynomial F and its inverse F^{-1} are EA-inequivalent. In this paper we continue the study of o-equivalence. We study a group of order 24 of transformations preserving o-polynomials which has been studied by Cherowitzo 25 years ago. We point out that three of the transformations he included in the group are not correct. We also deduce two more transformations preserving o-equivalence but providing potentially EA-inequivalent bent functions. We exhibit examples of infinite classes of o-polynomials for which at least three EA-inequivalent Niho bent functions can be derived.

Keywords: Bent function · Boolean function · Maximum nonlinearity · Niho bent function · o-polynomials · Walsh transform

1 Introduction

Boolean functions of n variables are binary functions over the vector space $\mathbb{F}_2^n$ of all binary vectors of length n, and can be viewed as functions over the Galois field $\mathbb{F}_{2^n}$, thanks to the choice of a basis of $\mathbb{F}_{2^n}$ over $\mathbb{F}_2$. In this paper, we shall always have this last viewpoint. Boolean functions are used in the pseudo-random generators of stream ciphers and play a central role in their security.

Bent functions, introduced by Rothaus [1] in 1976, are Boolean functions of an even number of variables n, that are maximally nonlinear in the sense that their nonlinearity, the minimum Hamming distance to all affine functions, is optimal. This corresponds to the fact that their Walsh transform takes the values $\pm 2^{n/2}$, only. Bent functions have attracted a lot of research interest in mathematics because of their relation to difference sets and to designs, and in the applications of mathematics to computer science because of their relations to coding theory and cryptography. Despite their simple and natural definition, bent functions turned out to admit a very complicated structure in general.

© Springer International Publishing Switzerland 2015
Ç. Koç et al. (Eds.): WAIFI 2014, LNCS 9061, pp. 155–168, 2015.
DOI: 10.1007/978-3-319-16277-5_9

An important focus of research was then to find constructions of bent functions. Many methods are known and some of them allow explicit constructions. We distinguish between primary constructions giving bent functions from scratch and secondary ones building new bent functions from one or several given bent functions (in the same number of variables or in different ones).

Bent functions are often better viewed in their bivariate representation, in the form $f(x, y)$, where x and y belong to $\mathbb{F}_2^m$ or to $\mathbb{F}_{2^m}$, where $m = n/2$. This representation has led to the two general families of explicit bent functions which are the original Maiorana-McFarland [5] and the Partial Spreads ($\mathcal{PS}_{ap}$) classes (this latter class is included in the more general but less explicit $\mathcal{PS}$ class). Bent functions can also be viewed in their univariate form (see Sect. 2), expressed by means of the trace function over $\mathbb{F}_{2^n}$. Finding explicit bent functions in this trace representation happens to be often still more difficult than in the bivariate representation. Survey references containing information on explicit primary constructions of bent functions in their bivariate and univariate forms are [2,4]. It is well known that some of these explicit constructions belong to the Maiorana-McFarland class and to the $\mathcal{PS}_{ap}$ class. When, in the early 1970s, Dillon introduced in his thesis [6] the two above mentioned classes, he also introduced another one denoted by H, where bentness was proven under some conditions which were not obvious to achieve. This made class H an example of non-explicit construction: at that time, Dillon was able to exhibit only functions belonging, up to the affine equivalence, to the Maiorana-McFarland class.

It was observed in [7] that the class of the, so called, Niho bent functions (introduced in [8] by Dobbertin *et al.*) is, up to EA-equivalence, equal to Dillon's class H. Note that functions in class H are defined in their bivariate representation and Niho bent functions had originally a univariate form only. Three infinite families of Niho binomial bent functions were constructed in [8] and one of these constructions was later generalized by Leander and Kholosha [9] into a function with 2^r Niho exponents. Another class was also extended in [10]. In [11] it was proven that some of these infinite families of Niho bent functions are EA-inequivalent to any Maiorana-McFarland function which implied that classes H and Maiorana-McFarland are different up to EA-equivalence.

In the same paper [7], the second author and Mesnager showed that Niho bent functions define o-polynomials and, conversely, every o-polynomial defines a Niho bent function. They also discovered that for a given o-polynomial F, equivalence of o-polynomials can lead to two EA-inequivalent Niho bent functions corresponding to F and F^{-1}. In this paper we study the group of transformations of order 24 preserving the equivalence of o-polynomials from [18], and we observe that for a given o-polynomial F these transformations can lead to up to four EA-inequivalent Niho bent functions corresponding to F, F^{-1}, $(xF(x^{-1}))^{-1}$ and $\left(x + xF\left(\frac{x+1}{x}\right)\right)^{-1}$. Hence, application of the equivalence of o-polynomials can be considered as a construction method for new (up to EA-equivalence) Niho bent functions from the known ones. For a given o-polynomial F the explicit polynomial expressions of Niho bent functions corresponding to F^{-1}, $(xF(x^{-1}))^{-1}$ and $\left(x + xF\left(\frac{x+1}{x}\right)\right)^{-1}$ are in general not easy to obtain because of the presence of the compositional inverse operator. However, for some known cases of o-polynomials we give the explicit bivariate expressions of the constructed functions, from which the univariate expression can be directly deduced.

The paper is organized as follows. In Sect. 2, we fix our main notation and recall the necessary background. The main results are contained in Sect. 3. We study the group of order 24 of transformations preserving o-polynomials studied by Cherowitzo. We begin in Subsect. 3.1 with the subgroup related to the symmetric group S_3 and we exhibit one more transformation preserving o-equivalence but providing potentially EA-inequivalent bent functions. We exhibit examples of infinite classes of o-polynomials for which three EA-inequivalent bent functions can be derived. Then we study in Subsect. 3.2 the whole group V. We point out that three of the transformations included by Cherowitzo in the group are not correct. Moreover, we exhibit still one more transformation preserving o-equivalence but providing potentially EA-inequivalent bent functions. We show that this fourth transformation can produce a Niho function EA-inequivalent to the three other ones.

2   Notation and Preliminaries

For any set E, we denote $E \setminus \{0\}$ by E^*. Throughout the paper, let n be even and $n = 2m$.

2.1   Trace Representation, Boolean Functions in Univariate and Bivariate Forms

For any positive integer k and any r dividing k, the trace function $\mathrm{Tr}_r^k()$ is the mapping from $\mathbb{F}_{2^k}$ to $\mathbb{F}_{2^r}$ defined by

$$\mathrm{Tr}_r^k(x) := \sum_{i=0}^{\frac{k}{r}-1} x^{2^{ir}} = x + x^{2^r} + x^{2^{2r}} + \cdots + x^{2^{k-r}}.$$

In particular, the *absolute trace* over $\mathbb{F}_{2^k}$ is the function $\mathrm{Tr}_1^k(x) = \sum_{i=0}^{k-1} x^{2^i}$ (in what follows, we just use $\mathrm{Tr}_k()$ to denote the absolute trace). Recall that the trace function satisfies the transitivity property $\mathrm{Tr}_k = \mathrm{Tr}_r \circ \mathrm{Tr}_r^k$.

The univariate representation of a Boolean function is defined as follows: we identify $\mathbb{F}_2^n$ (the n-dimensional vector space over $\mathbb{F}_2$) with $\mathbb{F}_{2^n}$ and consider the arguments of f as elements in $\mathbb{F}_{2^n}$. An inner product in $\mathbb{F}_{2^n}$ is $x \cdot y = \mathrm{Tr}_n(xy)$. There exists a unique univariate polynomial $\sum_{i=0}^{2^n-1} a_i x^i$ over $\mathbb{F}_{2^n}$ that represents f (this is true for any vectorial function from $\mathbb{F}_{2^n}$ to itself and therefore for any Boolean function since $\mathbb{F}_2$ is a subfield of $\mathbb{F}_{2^n}$). The algebraic degree of f (whose definition is related to the other representation of Boolean functions called algebraic normal form, see [2]) is equal in the univariate representation to the maximum 2-weight of the exponents of those monomials with nonzero coefficients, where the 2-weight $w_2(i)$ of an integer i is the number of ones in its binary expansion. Moreover, f being Boolean, its univariate representation can be written uniquely in the form of

$$f(x) = \sum_{j \in \Gamma_n} \mathrm{Tr}_{o(j)}(a_j x^j) + a_{2^n-1} x^{2^n-1},$$

where Γ_n is the set of integers obtained by choosing the smallest element in each cyclotomic coset modulo $2^n - 1$ (with respect to 2), $o(j)$ is the size of the cyclotomic coset containing j, $a_j \in \mathbb{F}_{2^{o(j)}}$ and $a_{2^n-1} \in \mathbb{F}_2$. The function f can also be written in a non-unique way as $\mathrm{Tr}_n(P(x))$ where $P(x)$ is a polynomial over $\mathbb{F}_{2^n}$.

The bivariate representation of a Boolean function is defined as follows: we identify $\mathbb{F}_2^n$ with $\mathbb{F}_{2^m} \times \mathbb{F}_{2^m}$ and consider the argument of f as an ordered pair (x, y) of elements in $\mathbb{F}_{2^m}$. There exists a unique bivariate polynomial $\sum_{0 \le i,j \le 2^m-1} a_{i,j} x^i y^j$ over $\mathbb{F}_{2^m}$ that represents f. The algebraic degree of f is equal to $\max_{(i,j) \mid a_{i,j} \ne 0}(w_2(i) + w_2(j))$. And f being Boolean, its bivariate representation can be written in the form $f(x, y) = \mathrm{Tr}_m(P(x, y))$, where $P(x, y)$ is some polynomial of two variables over $\mathbb{F}_{2^m}$.

2.2 Walsh Transform and Bent Functions

Let f be an n-variable Boolean function. Its *"sign"function* is the integer-valued function $\chi_f := (-1)^f$. The *Walsh transform* of f is the discrete Fourier transform of χ_f whose value at point $w \in \mathbb{F}_{2^n}$ is defined by

$$\widehat{\chi}_f(w) = \sum_{x \in \mathbb{F}_{2^n}} (-1)^{f(x) + \mathrm{Tr}_n(wx)}.$$

Definition 1. *For even n, a Boolean function f in n variables is said to be* bent *if for any $w \in \mathbb{F}_{2^n}$ we have $\widehat{\chi}_f(w) = \pm 2^{\frac{n}{2}}$.*

It is well known (see, for instance, [2]) that the algebraic degree of a bent Boolean function in $n > 2$ variables is at most $\frac{n}{2}$. This means that in the univariate representation of a bent function, all exponents i whose 2-weight is larger than m have zero coefficients a_i.

Bentness and algebraic degree (when larger than 1) are preserved by extended-affine (EA) equivalence. Two Boolean functions f and g are called *EA-equivalent* if there exists an affine automorphism A and an affine Boolean function ℓ such that $f = g \circ A + \ell$. There exists in the case of vectorial functions a more general notion of equivalence, called CCZ-equivalence, but for Boolean functions, it reduces to EA-equivalence, see [3].

2.3 Niho Bent Functions

A positive integer d (always understood modulo $2^n - 1$ with $n = 2m$) is a *Niho exponent* and $t \to t^d$ is a *Niho power function* if the restriction of t^d to $\mathbb{F}_{2^m}$ is linear or, equivalently, if $d \equiv 2^j \pmod{2^m - 1}$ for some $j < n$. As we consider

$\text{Tr}_n(at^d)$ with $a \in \mathbb{F}_{2^n}$, without loss of generality, we can assume that d is in the normalized form, i.e., with $j = 0$. Then we have a unique representation $d = (2^m - 1)s + 1$ with $2 \le s \le 2^m$. If some s is written as a fraction, this has to be interpreted modulo $2^m + 1$ (e.g., $1/2 = 2^{m-1} + 1$). Following are examples of bent functions consisting of one or more Niho exponents:

1. Quadratic functions $\text{Tr}_m(at^{2^m+1})$ with $a \in \mathbb{F}_{2^m}^*$ (here $s = 2^{m-1} + 1$).
2. Binomials of the form $f(t) = \text{Tr}_n(\alpha_1 t^{d_1} + \alpha_2 t^{d_2})$, where $2d_1 \equiv 2^m + 1$ (mod $2^n - 1$) and $\alpha_1, \alpha_2 \in \mathbb{F}_{2^n}^*$ are such that $(\alpha_1 + \alpha_1^{2^m})^2 = \alpha_2^{2^m+1}$. Equivalently, denoting $a = (\alpha_1 + \alpha_1^{2^m})^2$ and $b = \alpha_2$ we have $a = b^{2^m+1} \in \mathbb{F}_{2^m}^*$ and

$$f(t) = \text{Tr}_m(at^{2^m+1}) + \text{Tr}_n(bt^{d_2}).$$

We note that if $b = 0$ and $a \ne 0$ then f is a bent function listed under number 1. The possible values of d_2 are [8,10]:

$$d_2 = (2^m - 1)3 + 1,$$
$$6d_2 = (2^m - 1) + 6 \text{ (taking } m \text{ even)}.$$

These functions have algebraic degree m and do not belong to the completed Maiorana-McFarland class [11].
3. References [9,13] Take $1 < r < m$ with $\gcd(r, m) = 1$ and define

$$f(t) = \text{Tr}_n\left(a^2 t^{2^m+1} + (a + a^{2^m}) \sum_{i=1}^{2^{r-1}-1} t^{d_i} \right), \tag{1}$$

where $2^r d_i = (2^m - 1)i + 2^r$ and $a \in \mathbb{F}_{2^n}$ is such that $a + a^{2^m} \ne 0$. This function has algebraic degree $r + 1$ (see [12]) and belongs to the completed Maiorana-McFarland class [14].
4. Due to a discovered connection between Niho bent functions and o-polynomials we have some cases in bivariate representation (see Sect. 2.4).

Consider the above listed two binomial bent functions. If $\gcd(d_2, 2^n - 1) = d$ and $b = \beta^d$ for some $\beta \in \mathbb{F}_{2^n}$ then b can be "absorbed" in the power term t^{d_2} by a linear substitution of the variable t. In this case, up to EA-equivalence, $b = a = 1$. In particular, this applies to any b when $\gcd(d_2, 2^n - 1) = 1$ that holds in both cases except when $d_2 = (2^m - 1)3 + 1$ with $m \equiv 2$ (mod 4) where $d = 5$. In this exceptional case, we can get up to 5 different classes but the exact situation has to be further investigated.

2.4   Class $\mathcal{H}$ of Bent Functions and o-polynomials

In his thesis [6], Dillon introduced the class of bent functions denoted by H. The functions in this class are defined in their bivariate form as

$$f(x, y) = \text{Tr}_m(y + xF(yx^{2^m-2})), \tag{2}$$

where $x, y \in \mathbb{F}_{2^m}$ and F is a permutation of $\mathbb{F}_{2^m}$ such that $F(x)+x$ does not vanish and for any $\beta \in \mathbb{F}_{2^m}^*$, the function $F(x)+\beta x$ is 2-to-1 (i.e., the pre-image of any element of $\mathbb{F}_{2^m}$ is either a pair or the empty set). Dillon was just able to exhibit bent functions in H that also belong to the completed Maiorana-McFarland class. As observed by the second author and Mesnager [7, Proposition 1], this class can be slightly modified into the EA-equivalent class $\mathcal{H}$ defined as the set of (bent) functions g satisfying

$$g(x,y) = \begin{cases} \mathrm{Tr}_m \left(xG \left(\frac{y}{x}\right)\right), & \text{if } x \neq 0 \\ \mathrm{Tr}_m(\mu y), & \text{if } x = 0, \end{cases} \tag{3}$$

where $\mu \in \mathbb{F}_{2^m}$ and G is a mapping from $\mathbb{F}_{2^m}$ to itself satisfying the following necessary and sufficient conditions:

$$F : z \to G(z) + \mu z \text{ is a permutation on } \mathbb{F}_{2^m} \tag{4}$$

$$z \to F(z) + \beta z \text{ is 2-to-1 on } \mathbb{F}_{2^m} \text{ for any } \beta \in \mathbb{F}_{2^m}^*. \tag{5}$$

As proved in [7], condition (5) implies condition (4) and, thus, is necessary and sufficient for g being bent. Adding the linear term $\mathrm{Tr}_m((\mu+1)y)$ to (3) we obtain the original Dillon function (2). Therefore, functions in $\mathcal{H}$ and in the Dillon class are the same up to the addition of a linear term. It is observed in [7] that Niho bent functions are just functions in $\mathcal{H}$ in the univariate representation.

Any mapping F on $\mathbb{F}_{2^m}$ that satisfies (5) is called an *o-polynomial* [7]. Using (3), every o-polynomial results in a bent function in class $\mathcal{H}$. We list below the known o-polynomials due to Frobenius, Segre (1962), Glynn (1983), Cherowitzo (1998) and Payne (1985). The remaining two known cases (Subiaco and Adelaide) are listed in [7].

1. $F(z) = z^{2^i}$ with $\gcd(i, m) = 1$.
2. $F(z) = z^6$ with m odd.
3. $F(z) = z^{3 \cdot 2^k + 4}$ with $m = 2k - 1$.
4. $F(z) = z^{2^k + 2^{2k}}$ with $m = 4k - 1$.
5. $F(z) = z^{2^{2k+1} + 2^{3k+1}}$ with $m = 4k + 1$.
6. $F(z) = z^{2^k} + z^{2^k+2} + z^{3 \cdot 2^k + 4}$ with $m = 2k - 1$.
7. $F(z) = z^{\frac{1}{6}} + z^{\frac{1}{2}} + z^{\frac{5}{6}}$ with m odd.

In particular, functions (1) are obtained from the Frobenius map $z^{2^{m-r}}$ [14], binomial Niho bent functions with $d_2 = (2^m - 1)3 + 1$ correspond to Subiaco hyperovals [10] and functions with $6d_2 = (2^m - 1) + 6$ correspond to Adelaide hyperovals.

It was conjectured by Glynn [15] that all o-monomials are classified. This conjecture has already been proved for quadratic and cubic o-monomials [16,17].

There is a one-to one correspondence between o-polynomials satisfying the conditions $F(0) = 0$ and $F(1) = 1$ and hyperovals containing the "Fundamental Quadrangle" (i.e., the points with coordinates $(1, 0, 0)$, $(0, 1, 0)$, $(0, 0, 1)$ and $(1, 1, 1)$). Recall that a *hyperoval* of the projective plane $PG(2, 2^m)$ is a set of

$2^m + 2$ points no three of which are collinear. Two hyperovals are called equivalent if they are mapped to each other by a collineation (a permutation of the point set of $\mathrm{PG}(2, 2^m)$ mapping lines to lines). By the Fundamental Theorem of Projective Geometry, every hyperoval is equivalent to one containing the "Fundamental Quadrangle". If F is an o-polynomial satisfying $F(0) = 0$ and $F(1) = 1$ then the set

$$\Omega = \{(x, F(x), 1) | x \in \mathbb{F}_{2^m}\} \cup \{(1, 0, 0), (0, 1, 0)\}$$

is a hyperoval containing the "Fundamental Quadrangle". And vice versa, if Ω is a hyperoval of $\mathrm{PG}(2, 2^m)$ containing the "Fundamental Quadrangle" then it can be presented as a set

$$\Omega = \{(x, F(x), 1) | x \in \mathbb{F}_{2^m}\} \cup \{(1, 0, 0), (0, 1, 0)\}$$

for some o-polynomial F satisfying the conditions $F(0) = 0$ and $F(1) = 1$. For more information about hyperovals see for instance [17, 18].

Note that every o-polynomial F can be easily transformed into an o-polynomial $F^\alpha = \frac{F(x) + F(0)}{F(1) + F(0)}$ which satisfies $F^\alpha(0) = 0$ and $F^\alpha(1) = 1$. We call o-polynomials F_1 and F_2 *projectively equivalent* (or equivalent) if F_1^α and F_2^α define equivalent hyperovals. Note that the Niho bent function $g^\alpha(x, y)$ corresponding to F^α is EA-equivalent to the one defined by F. Indeed, g can be obtained from g^α by changing the variables x and y to $\frac{x}{F(1) + F(0)}$ and $\frac{y}{F(1) + F(0)}$, respectively, and adding a constant $\mathrm{Tr}_m\left(\frac{F(0)}{F(1) + F(0)}\right)$. We say that Niho bent functions are *o-equivalent* if they define projectively equivalent o-polynomials. As shown in [7], o-equivalent Niho bent functions may be EA-inequivalent. For example, Niho bent functions defined by o-polynomials F and F^{-1} are o-equivalent but they are, in general, EA-inequivalent. In the next section, we try to find transformations of o-polynomials which preserve projective equivalence but produce EA-inequivalent Niho bent functions.

3   EA-Equivalence Versus o-Equivalence for Niho Bent Functions

In this section, we study the action of a group V of transformations preserving o-polynomials, whose order is 24, and which is directly related to a group acting on hyperovals. We begin by studying in Subsect. 3.1 the subgroup of order 6 corresponding to the symmetric group S_3. Then, in Subsect. 3.2, we study the whole group V.

3.1   A Subgroup of Transformations Related to the Symmetric Group S_3

S_3 acts on the projective plane and leaves the set of o-polynomials invariant, see [18]. Hence, if $F(x)$ is an o-polynomial then the following list of images of

$(x, F(x), 1)$ under the action of S_3 (starting with the identity function)implicitly defines o-polynomials equivalent to F. Moreover, we know that two of the three transpositions generate S_3 by composition; we have chosen the transpositions numbered 3 and 6 below to express the other permutations, since they correspond to well identified transformations on o-polynomials (3 is the compositional inverse and 6 corresponds for the associated bent function in bivariate form to the swap between its arguments x and y as observed in [7]):

(1) $(x, F(x), 1)$;
(2) $(x, 1, F(x)) = 3 \circ 6 \circ 3 = 6 \circ 3 \circ 6$ (transposition);
(3) $(F(x), x, 1)$ (transposition);
(4) $(1, x, F(x)) = 6 \circ 3$;
(5) $(F(x), 1, x) = 3 \circ 6$;
(6) $(1, F(x), x)$ (transposition).

The following auxiliary observations in this section are easily obtained.

Observation 1. *All the transformations of o-polynomials corresponding to the symmetric group S_3 are compositions of transformations* 3 *and* 6.

Denoting by F^{-1} the compositional inverse of F (viewed as a mapping), by x^{inv} the multiplicative inverse x^{2^m-2} of x, and by $F'(x)$ the polynomial obtained from $F(x)$ by the action of 6, the o-polynomials obtained from $F(x)$ by transformations 1–6 are the following:

(1) $F(x)$;
(2) $\left(xF^{-1}(x^{inv})\right)^{-1} = ((F^{-1})')^{-1}(x)$;
(3) $F^{-1}(x)$;
(4) $xF^{-1}(x^{inv}) = (F^{-1})'(x)$;
(5) $\left(xF(x^{inv})\right)^{-1} = (F')^{-1}(x)$;
(6) $xF(x^{inv}) = F'(x)$.

Indeed, the action of 3 transforms $(x, F(x), 1)$ into $(F(x), x, 1)$ and the o-polynomial obtained from $F(x)$ by the action of 3 is then the compositional inverse of $F(x)$; the action of 6 transforms $(x, F(x), 1)$ into $(1, F(x), x) \equiv (x^{inv}, x^{inv}F(x), 1)$ and therefore the o-polynomial $F'(x)$ obtained from $F(x)$ by the action of 6 is such that $F'(x^{inv}) = x^{inv}F(x)$, which means that

$$F'(x) = xF(x^{inv});$$

the other expressions are deduced by composing these two transformations according to the expressions of 2, 4 and 5 by means of 3 and 6.

Proposition 1. *Applying the symmetric group S_3 to an o-polynomial F, one can derive up to three EA-inequivalent Niho bent functions corresponding to F, F^{-1} and $(F')^{-1}$.*

Proof. As noticed in [7], equivalent o-polynomials F and F^{-1} may define EA-inequivalent Niho bent functions while Niho bent functions defined by F and F' are always EA-equivalent (as already recalled, they correspond to each other by swapping x and y in the bivariate representation). Note that the equality $3 \circ 6 \circ 3 = 6 \circ 3 \circ 6$ for transformation 2 implies

$$((F^{-1})')^{-1} = ((F')^{-1})'.$$

Hence, Niho bent functions arising from $(F')^{-1}$ and $((F^{-1})')^{-1}$ are EA-equivalent. The former function corresponds to (5) and the latter to (2). We deduce that (1), (2), (3) may be EA-inequivalent but (4) is EA-equivalent to (3), (5) is EA-equivalent to (2) and (6) is EA-equivalent to (1). □

It can be easily observed that for an o-monomial $F(x) = x^d$ the algebraic degree of the corresponding Niho bent function equals $w_2(d) + w_2(2^m - d) = w_2(d) + m - w_2(d - 1)$, since the bivariate expression of this Niho bent function (with $\mu = 0$) is $g(x, y) = \mathrm{Tr}_m(x^{1-d}y^d)$.

In Table 1, we present exponents for known o-monomials $F_1(x) = x^d$, and we also display $F_2(x) = F_1^{-1}(x) = x^{\frac{1}{d}}$ and $F_3(x) = (F_2')^{-1}(x) = ((F_1^{-1})')^{-1}(x) = x^{\frac{1}{1-\frac{1}{d}}} = x^{\frac{d}{d-1}}$; we denote by d_i, $i = 1, 2, 3$, the algebraic degree of a Niho bent function obtained from F_i.

Since two EA-equivalent functions have the same algebraic degree, one can easily make conclusions on EA-inequivalence of many of the Niho bent functions arising from known o-monomials using data in Table 1. In particular, we can see that, in some cases, equivalent o-polynomials F, F^{-1} and $(F')^{-1}$ define three EA-inequivalent Niho bent functions. Note that F, $(F')^{-1}$ and F^{-1} correspond up to EA-equivalence to transformations 1, 2 and 3.

Proposition 2. *Applying the symmetric group S_3 to an o-monomial F, one can produce three surely EA-inequivalent Niho bent functions F, $(F')^{-1}$ and F^{-1} when F is either the first or the fifth o-monomial or the third o-monomial with k odd in Table 1.*

When F is the first o-monomial in Table 1, the bivariate polynomial expressions of the Niho bent function corresponding to $(F')^{-1}$ is $f(x, y) = \mathrm{Tr}_m(x^d y^{1-d})$ over $\mathbb{F}_{2^m}^2$ with

$$d = \sum_{i=1}^{h} 2^{si}, \quad h = s^{-1} \mod m, \ (s, m) = 1.$$

When F is the fifth o-monomial then the Niho bent function corresponding to $(F')^{-1}$ is $f(x, y) = \mathrm{Tr}_m(x^d y^{1-d})$ with $m = 4k + 1$ and

$$d = 2^{k+1} + \sum_{i=\frac{k+1}{2}}^{\frac{3k-1}{2}} 2^{2i+1} + \sum_{i=\frac{3k+1}{2}}^{2k} 2^{2i}, \quad \text{for } k \text{ odd};$$

$$d = 2 + \sum_{i=1}^{\frac{k}{2}} 2^{2i} + \sum_{i=\frac{k}{2}}^{\frac{3k-2}{2}} 2^{2i+1}, \quad \text{for } k \text{ even}.$$

Table 1. Niho bent functions from equivalent o-monomials

m	F_1	d_1	F_2	d_2	F_3	d_3
$(s,m)=1$	2^s	$m-s+1$	2^{m-s}	$s+1$	$\sum_{i=1}^h 2^{si},\, h=s^{-1} \bmod m$	m
$4k+1$	6	m	$\sum_{i=0}^{\frac{m-3}{2}} 2^{2i+1}+2^{m-1}$	m	$2+\sum_{i=1}^k 2^{4i}+\sum_{i=1}^k 2^{4i-1}$	m
$4k+3$					$4+\sum_{i=1}^k 2^{4i}+\sum_{i=1}^k 2^{4i+1}$	$m-1$
$2k-1$	$3\cdot 2^k+4$	$m-1$	$3\cdot 2^{k-1}-2$	m	$2^k+\sum_{i=\frac{k+1}{2}}^{k-1} 2^{2i}$	k, for k odd
					$2+\sum_{i=1}^{\frac{k-2}{2}} 2^{2i}$	m, for $k>2$ even
$4k-1$	$2^{2k}+2^k$	$3k$	$2^m-2^{3k-1}+2^{2k}-2^k$	$3k$	$2+\sum_{i=1}^{\frac{k-1}{2}} 2^{2i}+\sum_{i=\frac{k-1}{2}}^{\frac{3k-3}{2}} 2^{2i+1}$	m, for $k>1$ odd
					$2^k+\sum_{i=\frac{3k-2}{2}}^{} 2^{2i+1}+\sum_{i=\frac{3k}{2}}^{2k-1} 2^{2i}$	$3k$, for $k>0$ even
$4k+1$	$2^{3k+1}+2^{2k+1}$	$2k+1$	$2^m-2^{3k+1}+2^{2k+1}-2^k$	$3k+2$	$2^{k+1}+\sum_{i=\frac{k+1}{2}}^{\frac{3k-1}{2}} 2^{2i+1}+\sum_{i=\frac{3k}{2}}^{2k} 2^{2i}$	$3k+1$, for k odd
					$2+\sum_{i=1}^{\frac{k}{2}} 2^{2i}+\sum_{i=\frac{k}{2}}^{\frac{3k-2}{2}} 2^{2i+1}$	m, for k even

When F is the third o-monomial then the Niho bent function corresponding to $(F')^{-1}$ is $f(x,y)=\mathrm{Tr}_m(x^d y^{1-d})$ with $m=2k-1$, k odd and

$$d = 2^{k+1} + \sum_{i=\frac{k+1}{2}}^{\frac{3k-1}{2}} 2^{2i+1} + \sum_{i=\frac{3k+1}{2}}^{2k} 2^{2i}.$$

Note that one can get the univariate representation of these Niho bent functions by replacing in their bivariate representations $x = t+t^{2^m}$ and $y = at + (at)^{2^m}$ where a is primitive in $\mathbb{F}_{2^{2m}}$.

For those cases of Table 1 which have the same algebraic degree it is difficult to say whether the corresponding Niho bent functions are EA-inequivalent since computationally we can check it only for $m \leq 5$ which is very small.

Proposition 3. *Applying the symmetric group S_3 to the seventh o-polynomial one can derive at most two EA-inequivalent Niho bent functions.*

Proof. For the o-trinomial $F(z) = z^{\frac{1}{6}} + z^{\frac{1}{2}} + z^{\frac{5}{6}}$ (with m odd) we get $F' = F$ and, therefore, $(F')^{-1}$ equals F^{-1}. As observed in [7], $F^{-1} = (D_{\frac{1}{5}})^6$ (see below the definition of the Dickson polynomial $D_{\frac{1}{5}}$). □

We briefly recall the definition and the properties of Dickson polynomials. Every element x of $\mathbb{F}_{2^m}$ can be expressed uniquely in the form $h + \frac{1}{h}$ where $h \in \mathbb{F}_{2^n}^*$, since, for $x \neq 0$, the equation $h + \frac{1}{h} = x$ in the variable h is equivalent to $\left(\frac{h}{x}\right)^2 + \frac{h}{x} = \frac{1}{x^2}$ and has then two solutions inverses of each other, because the trace $\mathrm{Tr}_n\left(\frac{1}{x^2}\right)$ is null. This allows for every positive integer d to define the Dickson polynomial $D_d(X)$ over $\mathbb{F}_{2^m}$, such that $D_d\left(h + \frac{1}{h}\right) = h^d + \frac{1}{h^d}$ for every $h \in \mathbb{F}_{2^n}$. We have $D_0(X) = 0$, $D_1(X) = X$ and $D_{d+2}(X) = XD_{d+1}(X) + D_d(X)$ for every d (which implies that $D_d(\mathbb{F}_{2^m}) \subseteq \mathbb{F}_{2^m}$). Hence $D_5(X) = X + X^3 + X^5$. Moreover, D_d is a permutation polynomial if d is co-prime with $2^n - 1$ and since $D_d \circ D_{d'} = D_{dd'}$, the inverse of D_d is $D_{d'}$ where d' is the inverse of d or the inverse of $-d$ modulo $2^n - 1$. Denoting the inverse of 5 modulo $2^n - 1$ by $\frac{1}{5}$ we get $D_5^{-1} = D_{\frac{1}{5}}$.

3.2 The Group V of Order 24

In [18] Cherowitzo shows that S_3 can be extended to a group V of transformations whose order is 24 and which still leaves the set of o-polynomials invariant. This group can be obtained by applying S_3 to the following 4 transformations:

(a) $(x, F(x), 1)$;
(b) $(x + 1, F(x) + 1, 1)$ involutive;
(c) $(x, x + F(x), x + 1) = 6 \circ b \circ 6$ involutive;
(d) $(x + F(x), F(x), F(x) + 1) = 3 \circ 6 \circ b \circ 6 \circ 3$ involutive.

That is, all 24 transformations can be obtained from transformations a, b, c, d by permuting the coordinates. Note that when we write that $c = 6 \circ b \circ 6$, we do not apply b to $(1, F(x), x)$ but to the equivalent triple $(\frac{1}{x}, \frac{F(x)}{x}, 1)$. In fact, this is more easily seen by considering, instead of triples $(x, F(x), 1)$, equivalent classes of triples (u, v, w) under scalar multiplication; then 6 is viewed as $(u, v, w) \mapsto (w, v, u)$ and b is viewed as $(u, v, w) \mapsto (u + w, v + w, w)$. Similarly, $6 \circ b \circ 6$ corresponds to the sequence of transformations: $(u, v, w) \rightarrow (w, v, u) \rightarrow (w + u, v + u, u) \rightarrow (u, u + v, u + w)$ which is c and $3 \circ 6 \circ b \circ 6 \circ 3$ corresponds to $(u, v, w) \rightarrow (v, u, w) \rightarrow (w, u, v) \rightarrow (w + v, v + u, v) \rightarrow (v, u + v, v + w) \rightarrow (u + v, v, w + v)$ which is d.

Observation 2. *Transformations 3, 6 and b are generators of the group V.*

The o-polynomials obtained from F by transformations a, b, c, d are the following:

(a) $F(x)$;
(b) $F(x + 1) + 1$;
(c) $x + (x + 1)F(x(x + 1)^{inv})$;
(d) $\left(x + (x + 1)F^{-1}(x(x + 1)^{inv})\right)^{-1}$.

Note that:

$$b \circ 3 = 3 \circ b, \quad c \circ 3 = 3 \circ d, \quad d \circ 3 = 3 \circ c, \quad b \circ 6 = 6 \circ c$$
$$c \circ 6 = 6 \circ b, \quad d \circ 6 = 6 \circ d, \quad c \circ b = 6 \circ c, \quad b \circ c = 6 \circ b$$

$$d \circ b = 2 \circ d = 3 \circ 6 \circ 3 \circ d, \quad b \circ d = 2 \circ b = 3 \circ 6 \circ 3 \circ b$$

Remark 1. One can easily observe from the equalities above that V is indeed a group. We would like to emphasize this since the original list of elements of V from [18] (published 26 years ago) contains three wrong transformations and this list does not have the structure of a group.

Theorem 1. *The group V gives at most four EA-inequivalent functions. For an o-polynomial F the four potentially EA-inequivalent Niho bent functions correspond to F, F^{-1}, $(F')^{-1}$ and $F^{\circ}(x) = \left(x + xF\left(\frac{x+1}{x}\right)\right)^{-1}$ obtained from F by transformation $5 \circ b$.*

Proof. The only transformations in S_3 which can give mutually EA-inequivalent Niho functions are $1, 3$ and either 2 or 5 (we shall choose between 2 and 5 depending on conveniences). Hence, when searching for those transformations in the group V which could provide EA-inequivalent functions we need to consider only the following nine transformations:

$$b \rightarrow F(x+1) + 1, \qquad\qquad 3 \circ d \rightarrow (F^{-1})^*$$
$$c \rightarrow x + (x+1)F\big(x(x+1)^{inv}\big) = F^*(x), \; 5 \circ b \rightarrow \left(x + xF\big((x+1)x^{inv}\big)\right)^{-1} = F^{\circ}(x)$$
$$d \rightarrow ((F^{-1})^*)^{-1}, \qquad\qquad 5 \circ c \rightarrow ((F^*)')^{-1}$$
$$3 \circ b \rightarrow F^{-1}(x+1) + 1, \qquad\qquad 2 \circ d \rightarrow (((F^{-1})^*)')^{-1}$$
$$3 \circ c \rightarrow (F^*)^{-1}.$$

Two Niho bent functions obtained from an o-polynomial F and from its transformation by b are EA-equivalent [7]. Transformation c also leads to an EA-equivalent Niho bent function. Indeed,

$$g^*(x,y) = \mathrm{Tr}_m(xF^*(y/x)) = \mathrm{Tr}_m(y + (x+y)F(y/(x+y))) = \mathrm{Tr}_m(y + zF(y/z))$$

where $z = x + y$. Besides, $3 \circ d$ leads to a Niho bent function EA-equivalent to that defined by F^{-1} (obtained by 3).

Now let us consider the o-polynomial $((F^*)')^{-1}$ obtained from F by transformation $5 \circ c$. Since

$$((F^*)')^{-1}(x) = (F'(x+1) + 1)^{-1} = (F')^{-1}(x+1) + 1,$$

$((F^*)')^{-1}$ and $(F')^{-1}$ define EA-equivalent bent functions. For the same reasons, the o-polynomial $(((F^{-1})^*)')^{-1}$ obtained from F by transformation $2 \circ d$ defines Niho bent functions EA-equivalent to the one defined by $((F^{-1})')^{-1}$, and therefore, also EA-equivalent to the one defined by $(F')^{-1}$.

Transformations d and $3 \circ c$ give Niho bent functions EA-equivalent to that obtained from $5 \circ b$. Indeed, $5 \circ b = b \circ 3 \circ c$ and, since b gives EA-equivalent functions, the functions obtained by $5 \circ b$ and $3 \circ c$ are EA-equivalent. Besides, $d = b \circ 2 \circ b$. We know that $2 \circ b$ and $5 \circ b$ give EA-equivalent functions, and, therefore, $d = b \circ 2 \circ b$ and $5 \circ b$ also give EA-equivalent functions. $\square$

Proposition 4. *There exists an o-polynomial F such that the corresponding o-polynomial F° provides a Niho bent function EA-inequivalent to those defined by F, F^{-1} and $(F')^{-1}$.*

Proof. The o-polynomial $F^{\circ}(x) = \left(x + xF\left(\frac{x+1}{x}\right)\right)^{-1}$ obtained from F by transformation $5 \circ b$ may lead to a Niho bent function EA-inequivalent to those derived from F, F^{-1} and $(F')^{-1}$. Indeed, for $F(x) = x^6$ with $m = 5$ all three Niho bent functions defined by F, F^{-1} and $(F')^{-1}$ are EA-equivalent and, as it was checked with a computer, they are EA-inequivalent to the Niho bent function derived from F°. □

Proposition 5. *The group of transformations V produces exactly three EA-inequivalent functions when applied to an o-monomial $F(x) = x^{2^i}$, that is, the ones corresponding to F, F^{-1} and $(F')^{-1}$.*

Proof. One can easily observe that for $F(x) = x^{2^i}$ we get $F^\circ(x) = x^{\frac{1}{1-2^i}} = (F')^{-1}$. □

Proposition 6. $F^\circ(x) = (D_{\frac{1}{5}}(x))^{inv}$ *for the o-polynomial $F(x) = x^6$.*

Proof. For $F(x) = x^6$ we get

$$F^\circ(x) = (x^{inv} + (x^3)^{inv} + (x^5)^{inv})^{-1} = (D_5(x^{inv}))^{-1} = (D_{\frac{1}{5}}(x))^{inv}.$$

□

Problem 1. Find representations of F^{-1}, $(F')^{-1}$ and F° for all known o-polynomials F (the cases when it is not known).

Problem 2. In the group of all transformations which leave the set of o-polynomials invariant find all which lead to EA-inequivalent Niho bent functions.

4   Conclusions

In this paper we studied the symmetric group S_3 and its extension, the group of transformations V of order 24 preserving the equivalence of o-polynomials from [18]. We observed that three transformations needed to be corrected. We showed that these transformations can lead to up to four EA-inequivalent functions. More precisely, we prove that

- Within the orbit generated by S_3, at most 3 out of 6 Niho bent functions are pairwise EA-inequivalent. The upper bound is tight. It is an equality, for example, for translation and Glynn II o-polynomials. There are examples (e.g. Payne o-polynomials) for which the upper bound is not attained.
- Within the orbit generated by V, at most 4 out of 24 Niho bent functions are pairwise EA-inequivalent. Examples of o-polynomials are given that produce (with no possible restriction to S_3) two or three EA-inequalent bent functions. We actually do not have an example where 4 is attained and leave it as an open question.

Hence, application of the equivalence of o-polynomials can be considered as a construction method for new (up to EA-equivalence) Niho bent functions from the known ones. Note that the group of all transformations which leaves the set of o-polynomials invariant is much larger than V and is still to be studied.

Acknowledgements. We are very grateful to Bill Cherowitzo for useful discussions. This research was supported by Norwegian Research Council. The research of the first author was also supported by Fondation Sciences Mathématiques de Paris.

References

1. Rothaus, O.S.: On "bent" functions. J. Combin. Theory Ser. A **20**(3), 300–305 (1976)
2. Carlet, C.: Boolean functions for cryptography and error-correcting codes. In: Crama, Y., Hammer, P.L. (eds.) Boolean Models and Methods in Mathematics, Computer Science, and Engineering. Encyclopedia of Mathematics and its Applications, vol. 134, ch. 8, pp. 257–397. Cambridge University Press, Cambridge (2010)
3. Budaghyan, L., Carlet, C.: CCZ-equivalence of single and multi output Boolean functions. In: Post-proceedings of the Conference Fq9. AMS Contemporary Math., vol. 518, pp. 43–54 (2010)
4. Kholosha, Λ., Pott, A.: Bent and related functions. In: Mullen, G.L., Panario, D. (eds.) Handbook of Finite Fields. Discrete Mathematics and its Applications, ch. 9.3, pp. 255–265. CRC Press, London (2013)
5. McFarland, R.L.: A family of difference sets in non-cyclic groups. J. Combin. Theory Ser. A **15**(1), 1–10 (1973)
6. Dillon, J. F.: Elementary Hadamard difference sets, Ph.D. dissertation, University of Maryland (1974)
7. Carlet, C., Mesnager, S.: On Dillon's class H of bent functions, Niho bent functions and o-polynomials. J. Combin. Theory Ser. A **118**(8), 2392–2410 (2011)
8. Dobbertin, H., Leander, G., Canteaut, A., Carlet, C., Felke, P., Gaborit, P.: Construction of bent functions via Niho power functions. J. Combin. Theory Ser. A **113**(5), 779–798 (2006)
9. Leander, G., Kholosha, A.: Bent functions with 2^r Niho exponents. IEEE Trans. Inf. Theory **52**(12), 5529–5532 (2006)
10. Helleseth, T., Kholosha, A., Mesnager, S.: Niho bent functions and Subiaco hyperovals. In: Lavrauw, M., Mullen, G.L., Nikova, S., Panario, D., Storme, L. (eds.) Theory and Applications of Finite Fields. Contemporary Mathematics, vol. 579, pp. 91–101. American Mathematical Society, Providence (2012)
11. Budaghyan, L., Carlet, C., Helleseth, T., Kholosha, A., Mesnager, S.: Further results on Niho bent functions. IEEE Trans. Inf. Theory **58**(11), 6979–6985 (2012)
12. Budaghyan, L., Kholosha, A., Carlet, C., Helleseth, T.: Niho bent functions from quadratic o-monomials. In: Proceedings of the 2014 IEEE International Symposium on Information Theory (2014)
13. Li, N., Helleseth, T., Kholosha, A., Tang, X.: On the Walsh transform of a class of functions from Niho exponents. IEEE Trans. Inf. Theory **59**(7), 4662–4667 (2013)
14. Carlet, C., Helleseth, T., Kholosha, A., Mesnager, S.: On the dual of bent functions with 2^r Niho exponents. In: Proceedings of the 2011 IEEE International Symposium on Information Theory, pp. 657–661. IEEE, July/August 2011
15. Glynn, D.: Two new sequences of ovals in finite Desarguesian planes of even order. Combinatorial Mathematics, Lecture Notes in Mathematics, vol. 1036, pp. 217–229 (1983)
16. Cherowitzo, W.E., Storme, L.: α-Flocks with oval herds and monomial hyperovals. Finite Fields Appl. **4**(2), 185–199 (1998)
17. Vis, T.L.: Monomial hyperovals in Desarguesian planes, Ph.D. dissertation, University of Colorado Denver (2010)
18. Cherowitzo, W.: Hyperovals in Desarguesian planes of even order. Ann. Discrete Math. **37**, 87–94 (1988)

Third Invited Talk

L-Polynomials of the Curve $y^{q^n} - y = \gamma x^{q^h+1} - \alpha$ over $\mathbb{F}_{q^m}$

Ferruh Özbudak[1,2]([✉]) and Zülfükar Saygı[3]

[1] Department of Mathematics, Middle East Technical University,
Dumlupınar Bul., No:1, 06800 Ankara, Turkey
[2] Institute of Applied Mathematics, Middle East Technical University,
Dumlupınar Bul., No:1, 06800 Ankara, Turkey
ozbudak@metu.edu.tr
[3] Department of Mathematics, TOBB University of Economics and Technology,
Söğütözü, 06530 Ankara, Turkey
zsaygi@etu.edu.tr

Abstract. Let χ be a smooth, geometrically irreducible and projective curve over a finite field $\mathbb{F}_q$ of odd characteristic. The L-polynomial $L_\chi(t)$ of χ determines the number of rational points of χ not only over $\mathbb{F}_q$ but also over $\mathbb{F}_{q^s}$ for any integer $s \geq 1$. In this paper we determine L-polynomials of a class of such curves over $\mathbb{F}_q$.

Keywords: Algebraic curves · L-polynomials · Rational points

1 Introduction

In this paper we consider the L-polynomials of a special class of algebraic curves over finite fields of odd characteristic. Algebraic curves over finite fields have various applications in coding theory, cryptography, quasi-random numbers and related areas (see, for example, [6,7,12,14]). For these applications it is important to know the number of rational points of the curve. Throughout this paper by a curve we mean a smooth, geometrically irreducible and a projective curve over a finite field of odd characteristic.

A natural next step after the number of rational points of a curve is its L-polynomial (see, Sect. 2 below for a definition). It encodes the information for the number of rational points over all extensions of the finite field. Computing L-polynomial of curves is a difficult problem with some relations to algorithmic number theory and cryptography (see, for example, [3,4]).

L-polynomials of curves also carry important information about the Jacobian variety of a curve. Note that the Jacobian variety of a curve is an abelian variety of dimension g, where g is the genus of the curve. Namely it can be considered as the characteristic polynomial of the Frobenius action on its Jacobian variety (see, [13]).

This paper is organized as follows: In Sect. 2 we provide some basic facts and some open problems on algebraic curves over finite fields. Then in Sect. 3 we

© Springer International Publishing Switzerland 2015
Ç. Koç et al. (Eds.): WAIFI 2014, LNCS 9061, pp. 171–183, 2015.
DOI: 10.1007/978-3-319-16277-5_10

give our result on the L-polynomials of a class of curves in some special cases. In Appendix we present the number of rational points of the special curves.

2   Preliminaries

In this section we will present some basic facts and definitions on algebraic curves over finite fields and function fields.

Let $\mathbb{F}_q$ be a finite fields having q elements. A function field $\mathcal{K}$ is just a finite extension of the field of univariate rational fractions over $\mathbb{F}_q$, that is, a finite extension of the rational function field $\mathbb{F}_q(x)$. Hence $\mathcal{K} = \mathbb{F}_q(x, y)$ such that y is a root of an irreducible polynomial $r(T) = r_0(x) + r_1(x)T + \cdots + r_{n-1}(x)T^{n-1} + T^n \in \mathbb{F}_q(x)[T]$. Multiplying by a common denominator we get an irreducible polynomial $h(x, y) = y^n + y^{n-1}h_{n-1}(x) + \cdots + h_1(x)y + h_0(x) \in \mathbb{F}_q[x, y]$. In fact it is absolutely irreducible, i.e. irreducible over $\overline{\mathbb{F}_q}[x, y]$ (hence geometrically irreducible). Recall that a valuation ring of the function field $\mathcal{K}$ over $\mathbb{F}_q$ is a ring $\mathcal{O}$ such that $\mathbb{F}_q \subset \mathcal{O} \subset \mathcal{K}$ and for any $z \in \mathcal{K}$, $z \in \mathcal{O}$ or $z^{-1} \in \mathcal{O}$. A place $\mathcal{P}$ of the function field $\mathcal{K}$ over $\mathbb{F}_q$ is the maximal ideal of some valuation ring $\mathcal{O}$ of $\mathcal{K}$. Let $\mathbb{P}_\mathcal{K}$ denote the set of all places of $\mathcal{K}$ over $\mathbb{F}_q$. We remark that if $\mathcal{O}$ is a valuation ring of $\mathcal{K}$ over $\mathbb{F}_q$ and $\mathcal{P}$ is its maximal ideal, then $\mathcal{O}$ can be uniquely determined by $\mathcal{P}$, that is, $\mathcal{O} = \mathcal{O}_\mathcal{P} = \{z \in \mathcal{K} \mid z^{-1} \notin \mathcal{P}\}$. Note that as $\mathcal{P}$ is a maximal ideal of $\mathcal{O}_\mathcal{P}$ then $\mathcal{O}_\mathcal{P}/\mathcal{P}$ is called the residue class field of $\mathcal{P}$. Then the extension degree of $\mathcal{O}_\mathcal{P}/\mathcal{P}$ over $\mathbb{F}_q$ is called the degree of the place $\mathcal{P}$, that is, $\deg \mathcal{P} = [\mathcal{O}_\mathcal{P}/\mathcal{P} : \mathbb{F}_q]$.

The formal sum $D = \sum\limits_{\mathcal{P} \in \mathbb{P}_\mathcal{K}} n_\mathcal{P}\mathcal{P}$ is called a divisor. Here note that $n_\mathcal{P}$ is zero for all except finitely many entries throughout the set $\mathbb{P}_\mathcal{K}$ and $n_\mathcal{P} \in \mathbb{Z}$. Then the degree of a divisor D is defined as $\deg D = \sum\limits_{\mathcal{P} \in \mathbb{P}_\mathcal{K}} n_\mathcal{P} \deg \mathcal{P}$. Let $D(\mathcal{K})$ be the set of all divisors of $\mathcal{K}$. We know that it is an infinite abelian group. Let $D^0(\mathcal{K})$ denote the subset consisting of divisors D of $\mathcal{K}$ with $\deg D = 0$. Note that $D^0(\mathcal{K})$ is a subgroup of $D(\mathcal{K})$. Let $\mathcal{P}(\mathcal{K})$ be the subset of $D^0(\mathcal{K})$ consisting of the principal divisors of $\mathcal{K}$. Then the factor group

$$Jac(\mathcal{K}) = D^0(\mathcal{K})/\mathcal{P}(\mathcal{K})$$

is called the Jacobian of $\mathcal{K}$.

For any divisors $D = \sum_{\mathcal{P} \in \mathbb{P}_\mathcal{K}} n_\mathcal{P}\mathcal{P}$ and $D' = \sum_{\mathcal{P} \in \mathbb{P}_\mathcal{K}} n'_\mathcal{P}\mathcal{P}$ in $D(\mathcal{K})$ we say $D \geq D'$ if $n_\mathcal{P} \geq n'_\mathcal{P}$ for all $\mathcal{P} \in \mathbb{P}_\mathcal{K}$. If $n_\mathcal{P} = 0$ for all $\mathcal{P} \in \mathbb{P}_\mathcal{K}$ then we call $D = 0$ as the zero divisor. For integers $n \geq 0$ let us define

$$A_n = |\{A \in D(\mathcal{K}) : A \geq 0 \text{ and } \deg A = 0\}|.$$

We know that A_n is a finite number. Then the zeta function $Z_\mathcal{K}(t)$ is defined as the power series

$$Z_\mathcal{K}(t) = \sum_{n=0}^{\infty} A_n t^n \in \mathbb{C}[[t]].$$

The theory of algebraic curves is essentially equivalent to the theory of function fields. For a brief survey of the relations between algebraic curves and function fields we refer to [12, Appendix B]. Now using the above zeta function we can define the L-polynomial of a curve χ (or the L-polynomial of the corresponding function field). The following is one of the most important results in the theory of algebraic curves over finite fields (see, for example, [12]).

Theorem 1. *Let χ be a curve over $\mathbb{F}_q$ with full constant field $\mathbb{F}_q$ and $F = \mathbb{F}_q(x, y)$ be its function field. Then*

1. *There exists $L_F(t) = 1 + a_1 t + a_2 t^2 + \cdots + a_{2g-1} t^{2g-1} + q^g t^{2g} \in \mathbb{Z}[t]$ such that*

$$Z_F(t) = \frac{L_F(t)}{(1-t)(1-qt)}.$$

   Here $L_F(t)$ is called the L-polynomial of χ (or F). Note that g is the genus of χ (or F).
2. *$a_{2g-i} = q^{g-i} a_i$ for $1 \leq i \leq g$.*
3. *$L_F(t) = \prod_{i=1}^{2g}(1 - \alpha_i t)$ with $\alpha_i \in \mathbb{C}$.*
4. *Hasse-Weil Theorem is equivalent to*

$$\log |\alpha_i| = \frac{1}{2} \text{ for } i = 1, 2, \ldots, 2g.$$

(Note that this is an analogue of the classical Riemann Hypothesis (which is still an open problem) in positive characteristic.)

For the curve χ over $\mathbb{F}_q$, let $F = \mathbb{F}_q(x, y)$ denote its function field and let $N(\chi)$ be the number of rational points of χ, i.e. the number of degree one places of F. Then Hasse-Weil Theorem gives

$$q + 1 - 2g\sqrt{q} \leq N(\chi) \leq q + 1 + 2g\sqrt{q}$$

where g is the genus of F. Serre improved this slightly to

$$q + 1 - g\lfloor 2q^{1/2} \rfloor \leq N(\chi) \leq q + 1 + g\lfloor 2q^{1/2} \rfloor$$

(for some q). We know that there exist curves satisfying the Hasse-Weil bounds. If the upper bound is attained then the curve is called a maximal curve and if the lower bound is attained then the curve is called a minimal curve. But in general, it is difficult to determine $N(\chi)$. Let $N(F.\mathbb{F}_{q^i})$ denote the number of degree one places of F over $\mathbb{F}_{q^i}$ for some i. In order to determine the L-polynomial of F, one needs to determine $N(F.\mathbb{F}_{q^i})$ for all extensions $\mathbb{F}_{q^i}/\mathbb{F}_q$ for $i = 1, 2, \ldots, g$. Then the L-polynomial $L_F(t) = 1 + a_1 t + \cdots + a_{2g-1} t^{2g-1} + q^g t^{2g} \in \mathbb{Z}[t]$ gives $N(F.\mathbb{F}_{q^i})$ for all $i \in \mathbb{Z}^+$. Moreover $L_F(t)$ gives also further "geometric" information. For example:

- $L_F(1)$ = the class number of the function field F (see, [12, Theorem 5.1.15]),
- p-rank (or Hasse-Witt invariant) of its Jacobian variety (see, [11]),
- Newton polygon of the curve (or its Jacobian) (see, [5]).

If we consider the projective line, that is $\chi = \mathbb{P}^1$ or $F = \mathbb{F}_q(x)$ then we know that it has genus 0. This gives that $L_F(t) = 1$ and $Z_F(t) = \frac{1}{(1-t)(1-q^t)}$. Furthermore, if we consider a maximal curve χ of genus g over $\mathbb{F}_q$ with $g \geq 1$ then q should be a square and $L_\chi(t) = (1 - \sqrt{q}t)^{2g}$ Also if χ is a minimal curve of genus g over $\mathbb{F}_q$ then again q should be a square and $L_\chi(t) = (1 + \sqrt{q}t)^{2g}$ Note that it is still an open problem (in general) if there exists a maximal (and minimal) curve of genus g over $\mathbb{F}_q$. For maximal curves of genus g over $\mathbb{F}_q$ we know that $g \leq \frac{(\sqrt{q}+1)\sqrt{q}}{2}$. Moreover, the Hermitian curve $H : y^q + y = x^{q+1}$ over $\mathbb{F}_{q^2}$ is the only maximal curve of genus $\frac{(q+1)q}{2}$ over $\mathbb{F}_{q^2}$ up to birational isomorphism (see, [10]).

Consider an elliptic curve E over $\mathbb{F}_q$. Then by Hasse-Weil theorem, for the number $N(E)$ of rational places of E we have $N(E) = q + 1 - b$ with $-2\sqrt{q} \leq -b \leq 2\sqrt{q}$. In general we do not know which values of the Hasse-Weil interval are attained. Here for $g = 1$ (elliptic curves), we know it exactly by the following result of Waterhouse [16] (see also [9,15]).

Theorem 2. *There exists an elliptic curve over $\mathbb{F}_{p^n}$ with $N(E) = p^n + 1 - b$ if and only if b satisfies one of the following conditions:*

1. $gcd(b, p) = 1$
2. n is even and $b = \pm 2\sqrt{q}$ (maximal, minimal)
3. n is even, $p \not\equiv 1 \mod 3$ and $b = \pm\sqrt{q}$
4. n is odd, $p \in \{2, 3\}$ and $b = \pm p^{n+1/2}$
5. n is odd and $b = 0$
6. n is even $p \not\equiv 1 \mod 4$ and $b = 0$.

Furthermore, $L_E(t) = 1 + a_1 t + qt^2$ where $a_1 = N(E) - (q+1) = -b$.

3   L-polynomials of a Class of Algebraic Curves

Let p be an odd prime. For positive integers e and m, let $q = p^e$ and let $\mathbb{F}_q$ and $\mathbb{F}_{q^m}$ denote the finite fields with q and q^m elements. Let n be a positive integer dividing m. Let $\mathrm{Tr}_{\mathbb{F}_{q^m}/\mathbb{F}_{q^n}}$ denote the relative trace map.

Let h be a nonnegative integer and $\alpha, \gamma \in \mathbb{F}_{q^m}$ with $\gamma \neq 0$. Let $N(m, n)$ denote the cardinality

$$N(m, n) = \left| \left\{ x \in \mathbb{F}_{q^m} \mid \mathrm{Tr}_{\mathbb{F}_{q^m}/\mathbb{F}_{q^n}} \left(\gamma x^{q^h + 1} - \alpha \right) = 0 \right\} \right|.$$

Let χ be the Artin-Schreier type curve given by

$$\chi : \quad y^{q^n} - y = \gamma x^{q^h + 1} - \alpha. \tag{1}$$

Note that the genus of the curve χ is $g(\chi) = \frac{(q^n - 1)q^h}{2}$ and for the number $N(\chi)$ of its $\mathbb{F}_{q^m}$-rational points using Hilbert's Theorem 90 we have

$$N(\chi) = 1 + q^n N(m, n),$$

and hence determining $N(\chi)$ is the same as determining $N(m, n)$. This number is determined exactly in many cases in [8] and we have included these results in the Appendix for the completeness of the paper. Using these results and some techniques in [12, Sect. 5], we have obtained the L-polynomial $L_\chi(t)$ of χ in some cases as follows.

Assume that q, m and n are given. Then to compute the L-polynomial $L_\chi(t)$ of χ we follow the following steps:

1. Compute $N(lm, n)$ for $l \in \mathbb{Z}^+$ using Theorems 4, 5 and 6.
2. Compute $S(lm, n) = q^n N(lm, n) - q^{lm}$.
3. Compute $\dfrac{L'_\chi(t)}{L_\chi(t)} = \sum\limits_{l=1}^{\infty} S(lm, n) t^{l-1}$ where $L'_\chi(t)$ is the derivative of $L_\chi(t)$.
4. Take the integral of both sides and then simplify the result to get $L_\chi(t)$.

Before presenting the numerical results on L-polynomials we want to remark the relation between the Hermitian curves and our curves of the form (1).

Remark 1. Here we remark that our curves in (1) covers the well known Hermitian curves of the form

$$y^q + y = x^{q+1} \text{ over } \mathbb{F}_{q^2}.$$

In (1) if we take $n = 1$, $h = 1$, $m = 2$ and $\alpha = 0$ and change the variable y as $y = cz$ for some nonzero $c \in \mathbb{F}_{q^2}$ then we have

$$(cz)^q - cz = \gamma x^{q+1}.$$

Dividing by c^q we get

$$z^q - \frac{1}{c^{q-1}} z = \frac{\gamma}{c^q} x^{q+1}.$$

Then we should have $\dfrac{1}{c^{q-1}} = -1$ and $\dfrac{\gamma}{c^q} = 1$. Let w be a generator of the multiplicative group $\mathbb{F}_{q^2} \setminus \{0\}$. Then by choosing $c = w^{\frac{q+1}{2}}$ and $\gamma = c^q = w^{\frac{q(q+1)}{2}}$ we obtain the desired result.

Following the above steps we have calculated some L-polynomials of the corresponding curves χ as follows.

Example 1. Let χ be the curve $y^q - y = x^{q+1}$ over $\mathbb{F}_{q^4}$ then the L-polynomial $L_\chi(t)$ becomes

$$L_\chi(t) = \left(1 - q^2 t\right)^{q(q-1)}.$$

Example 2. Let χ be the curve $y^q - y = x^{q+1}$ over $\mathbb{F}_{q^2}$ then the L-polynomial $L_\chi(t)$ becomes

$$L_\chi(t) = (1 - qt)^{\frac{q^2-1}{2}} (1 + qt)^{\frac{(q-1)^2}{2}}.$$

Example 3. Let χ be the curve $y^q - y = x^{q^2+1}$ over $\mathbb{F}_{q^4}$ then the L-polynomial $L_\chi(t)$ becomes

$$L_\chi(t) = \left(1 - q^2 t\right)^{\frac{(q^2+1)(q-1)}{2}} \left(1 + q^2 t\right)^{\frac{(q^2-1)(q-1)}{2}}.$$

Example 4. Let χ be the curve $y^q - y = x^{q^4+1}$ over $\mathbb{F}_{q^6}$ then the L-polynomial $L_\chi(t)$ becomes

$$L_\chi(t) = \begin{cases} \left(1 + q^3 t\right)^{q-1} \left(1 - q^{24} t^8\right)^{\frac{q^4(q-1)}{8}} & \text{if } q \equiv 3 \mod 4, \\ \left(1 - q^3 t\right)^{q-1} \left(1 - q^{24} t^8\right)^{\frac{q^4(q-1)}{8}} & \text{if } q \equiv 1 \mod 4. \end{cases}$$

Example 5. Let χ be the curve $y^q - y = x^{q^4+1}$ over $\mathbb{F}_{q^5}$ then the L-polynomial $L_\chi(t)$ becomes

$$L_\chi(t) = \begin{cases} \left(1 - q^5 t\right)^{q-1} \left(1 - q^{40} t^{16}\right)^{\frac{(q^4-1)(q-1)}{16}} & \text{if } q \equiv 3 \mod 4, \\ \dfrac{\left(1 - q^5 t^2\right)^{\frac{q-1}{2}} \left(1 - q^5 t\right)^{q-1} \left(1 - q^{40} t^{16}\right)^{\frac{(q^4-1)(q-1)}{16}}}{\left(1 + q^5 t^2\right)^{\frac{q-1}{2}}} & \text{if } q \equiv 1 \mod 4. \end{cases}$$

Example 6. Let χ be the curve $y^{q^2} - y = x^{q+1}$ over $\mathbb{F}_{q^4}$ then the L-polynomial $L_\chi(t)$ becomes

$$L_\chi(t) = \begin{cases} \left(1 - q^2 t\right)^{2(q^2-1)} \left(1 + q^2 t\right)^{q^2-1} & \text{if } q = 3, \\ \left(1 - q^2 t\right)^{q^2-1} \left(1 - q^6 t^3\right)^{\frac{4(q^2-1)}{3}} & \text{if } q = 5. \end{cases}$$

In the following result we see that the L-polynomial $L_\chi(t)$ depends on the factorization of $q + 1$ in some cases.

Theorem 3. *Let χ be the curve $y^{q^2} - y = x^{q+1}$ over $\mathbb{F}_{q^2}$ and assume that $q + 1 = 2^{\nu_0} \theta^{\nu_1}$ for some odd prime θ and positive integers ν_0, ν_1. Then the L-polynomial $L_\chi(t)$ becomes*

$$\left(\frac{1 - q^\theta t^\theta}{1 + q^\theta t^\theta}\right)^{\frac{q^2-1}{2\theta}} \left(\prod_{i=1}^{\nu_0} \left(1 - q^{2^i} t^{2^i}\right)^{\frac{q^2-1}{2}}\right) \left(\prod_{i=1}^{\nu_1} \left(1 - q^{2\theta^i} t^{2\theta^i}\right)^{\frac{(q^2-1)(\theta-1)}{\theta}}\right)$$

$$\left(\prod_{i=2}^{\nu_0} \prod_{j=1}^{\nu_1} \left(1 - q^{2^i \theta^j} t^{2^i \theta^j}\right)^{\frac{(q^2-1)(\theta-1)}{2\theta}}\right).$$

Proof. We have $m = 2$, $h = 1$ and $n = 2$. Then using Theorem 6 for any positive integer l we get

$$N(lm, n) = \begin{cases} q^{2l-2} & \text{if } l \text{ is odd and } \theta \nmid l, \\ q^{2l-2} - (q^2 - 1)q^{l-2} & \text{if } l \text{ is odd and } \theta \mid l, \\ q^{2l-2} + (1 - B_1)(q^2 - 1)q^{l-2} & \text{if } l \text{ is even} \end{cases}$$

where $B_1 = \gcd(l, q + 1)$. This gives

$$S(lm, n) = \begin{cases} 0 & \text{if } l \text{ is odd and } \theta \nmid l, \\ -(q^2 - 1)q^l & \text{if } l \text{ is odd and } \theta \mid l, \\ (1 - B_1)(q^2 - 1)q^l & \text{if } l \text{ is even} \end{cases} \tag{2}$$

where $B_1 = \gcd(l, q + 1)$. Then we need to compute

$$\frac{L'_\chi(t)}{L_\chi(t)} = \sum_{l=1}^{\infty} S(lm, n)t^{l-1}$$

$$= \sum_{l \text{ odd}} S(lm, n)t^{l-1} + \sum_{l \text{ even}} S(lm, n)t^{l-1}. \tag{3}$$

First we compute the sum $\sum_{l \text{ odd}} S(lm, n)t^{l-1}$ using (2).

$$\sum_{l \text{ odd}} S(lm, n)t^{l-1} = \sum_{l \text{ odd and } \theta \nmid l} S(lm, n)t^{l-1} + \sum_{l \text{ odd and } \theta \mid l} S(lm, n)t^{l-1}$$

$$= 0 + \sum_{l \text{ odd and } \theta \mid l} -(q^2 - 1)q^l t^{l-1}$$

$$= \sum_{k=0}^{\infty} -(q^2 - 1)q^{(2k+1)\theta}t^{(2k+1)\theta-1}$$

$$= -(q^2 - 1)\frac{q^\theta t^{\theta-1}}{1 - q^{2\theta}t^{2\theta}}. \tag{4}$$

Now we compute $\sum_{l \text{ even}} S(lm, n)t^{l-1}$ using (2), in short we will denote this sum by $\sum_{l \text{ even}}$. This sum depends on the value of $B_1 = \gcd(l, q + 1)$. As we assume that $q + 1 = 2^{\nu_0}\theta^{\nu_1}$ and l is even, then depending on the value of l we have $B_1 = \gcd(l, q + 1) = 2^i \theta^j$ with $1 \leq i \leq \nu_0$ and $0 \leq j \leq \nu_1$. So we can write

$$\sum_{l \text{ even}} = \sum_{l \text{ even and } B_1=2} + \sum_{l \text{ even and } B_1=4} + \cdots + \sum_{l \text{ even and } B_1=2^{\nu_0}\theta^{\nu_1-1}} + \sum_{l \text{ even and } B_1=2^{\nu_0}\theta^{\nu_1}}. \tag{5}$$

Note that $B_1 = 2$ if and only if $2 \mid l$, $4 \nmid l$ and $2\theta \nmid l$. Using this observation and the inclusion-exclusion principle we can write

$$\sum_{l \text{ even and } B_1=2} = \sum_{2\mid l} - \sum_{4\mid l} - \sum_{2\theta\mid l} + \sum_{4\theta\mid l}. \tag{6}$$

Furthermore, $B_1 = 2^i \theta^j$ with $1 \leq i \leq \nu_0 - 1$ and $0 \leq j \leq \nu_1 - 1$ if and only if $2^i \theta^j \mid l$, $2^{i+1}\theta^j \nmid l$ and $2^i \theta^{j+1} \nmid l$. Similarly, using these observations we can write

$$\sum_{l \text{ even and } B_1 = 2^i \theta^j} = \sum_{2^i \theta^j \mid l} - \sum_{2^{i+1}\theta^j \mid l} - \sum_{2^i \theta^{j+1} \mid l} + \sum_{2^{i+1}\theta^{j+1} \mid l}. \tag{7}$$

But if $B_1 = 2^i \theta^j$ with $i = \nu_0$ and $0 \leq j \leq \nu_1 - 1$ then we have

$$\sum_{l \text{ even and } B_1 = 2^{\nu_0} \theta^j} = \sum_{2^{\nu_0}\theta^j \mid l} - \sum_{2^{\nu_0}\theta^{j+1} \mid l} \tag{8}$$

and similarly if $B_1 = 2^i \theta^j$ with $1 \leq i \leq \nu_0 - 1$ and $j = \nu_1$ then we have

$$\sum_{l \text{ even and } B_1 = 2^i \theta^{\nu_1}} = \sum_{2^i \theta^{\nu_1} \mid l} - \sum_{2^{i+1}\theta^{\nu_1} \mid l}. \tag{9}$$

Note that, on the right hand side of (6) and (7) we have 4 summations, but on the right hand side of (8) and (9) we have only 2 summations. The only remaining case is $B_1 = 2^{\nu_0}\theta^{\nu_1}$. But in this case we have

$$\sum_{l \text{ even and } B_1 = 2^{\nu_0}\theta^{\nu_1}} = \sum_{2^{\nu_0}\theta^{\nu_1} \mid l}. \tag{10}$$

Therefore, to compute the sum in (5) we need to compute the sums on the right hand side of the Eqs. (6), (7), (8), (9) and (10).

If we concentrate on the right hand sides of the Eqs. (6), (7), (8), (9) and (10) we see that $\sum_{2 \mid l}$ occurs only 1 times in the case $B_1 = 2$, $\sum_{2^i \mid l}$ with $2 \leq i \leq \nu_0$ occurs 2 times in the cases $B_1 = 2^i$ and $B_1 = 2^{i-1}$, $\sum_{2\theta^j \mid l}$ with $1 \leq j \leq \nu_1$ occurs 2 times in the cases $B_1 = 2\theta^j$ and $B_1 = 2\theta^{j-1}$ and $\sum_{2^i \theta^j \mid l}$ with $2 \leq i \leq \nu_0$, $1 \leq j \leq \nu_1$ occurs 4 times in the cases $B_1 = 2^i \theta^j$, $B_1 = 2^{i-1}\theta^j$, $B_1 = 2^i \theta^{j-1}$ and $B_1 = 2^{i-1}\theta^{j-1}$. Using these observations and combining (5) with (6), (7), (8), (9) and (10) we obtain that

$$\sum_{l \text{ even}} = \sum_{2 \mid l \text{ and } B_1 = 2}$$

$$+ \sum_{i=2}^{\nu_0} \left(\sum_{2^i \mid l \text{ and } B_1 = 2^i} - \sum_{2^i \mid l \text{ and } B_1 = 2^{i-1}} \right)$$

$$+ \sum_{j=1}^{\nu_1} \left(\sum_{2\theta^j \mid l \text{ and } B_1 = 2\theta^j} - \sum_{2\theta^j \mid l \text{ and } B_1 = 2\theta^{j-1}} \right)$$

$$+ \sum_{i=2}^{\nu_0} \sum_{j=1}^{\nu_1} \left(\sum_{2^i \theta^j \mid l \text{ and } B_1 = 2^i \theta^j} - \sum_{2^i \theta^j \mid l \text{ and } B_1 = 2^{i-1}\theta^j} - \sum_{2^i \theta^j \mid l \text{ and } B_1 = 2^i \theta^{j-1}} \right.$$

$$\left. + \sum_{2^i \theta^j \mid l \text{ and } B_1 = 2^{i-1}\theta^{j-1}} \right).$$

Then using (2) we have

$$\sum_{2\mid l \text{ and } B_1=2} = \sum_{2\mid l} -(q^2-1)q^l t^{l-1}$$

$$= \sum_{k=1}^{\infty} -(q^2-1)q^{2k}t^{2k-1}$$

$$= -(q^2-1)\frac{q^2 t}{1-q^2 t^2}, \tag{11}$$

$$\sum_{i=2}^{\nu_0}\left(\sum_{2^i\mid l \text{ and } B_1=2^i} - \sum_{2^i\mid l \text{ and } B_1=2^{i-1}}\right) = \sum_{i=2}^{\nu_0}\left(\sum_{2^i\mid l}\left((1-2^i)-(1-2^{i-1})\right)(q^2-1)q^l t^{l-1}\right)$$

$$= \sum_{i=2}^{\nu_0}\left(\sum_{k=1}^{\infty} -2^{i-1}(q^2-1)q^{2^i k}t^{2^i k-1}\right)$$

$$= \sum_{i=2}^{\nu_0}\left(-2^{i-1}(q^2-1)\frac{q^{2^i}t^{2^i-1}}{1-q^{2^i}t^{2^i}}\right), \tag{12}$$

$$\sum_{j=1}^{\nu_1}\left(\sum_{2\theta^j\mid l \text{ and } B_1=2\theta^j} - \sum_{2\theta^j\mid l \text{ and } B_1=2\theta^{j-1}}\right) = \sum_{j=1}^{\nu_1}\left(\sum_{2\theta^j\mid l}\left((1-2\theta^j)-(1-2\theta^{j-1})\right)(q^2-1)q^l t^{l-1}\right)$$

$$= \sum_{j=1}^{\nu_1}\left(\sum_{k=1}^{\infty} -2\theta^{j-1}(\theta-1)(q^2-1)q^{2\theta^j k}t^{2\theta^j k-1}\right)$$

$$= \sum_{j=1}^{\nu_1}\left(-2\theta^{j-1}(\theta-1)(q^2-1)\frac{q^{2\theta^j}t^{2\theta^j-1}}{1-q^{2\theta^j}t^{2\theta^j}}\right), \tag{13}$$

$$\sum_{i=2}^{\nu_0}\sum_{j=1}^{\nu_1}\left(\sum_{2^i\theta^j\mid l \text{ and } B_1=2^i\theta^j} - \sum_{2^i\theta^j\mid l \text{ and } B_1=2^{i-1}\theta^j} - \sum_{2^i\theta^j\mid l \text{ and } B_1=2^i\theta^{j-1}}\right.$$

$$\left. + \sum_{2^i\theta^j\mid l \text{ and } B_1=2^{i-1}\theta^{j-1}}\right)$$

$$= \sum_{i=2}^{\nu_0}\sum_{j=1}^{\nu_1}\left(\sum_{2^i\theta^j\mid l}\left((1-2^i\theta^j)-(1-2^{i-1}\theta^j)-(1-2^i\theta^{j-1})+(1-2^{i-1}\theta^{j-1})\right)\right.$$

$$(q^2-1)q^l t^{l-1})$$

$$= \sum_{i=2}^{\nu_0}\sum_{j=1}^{\nu_1}\left(\sum_{k=1}^{\infty}\left(-2^{i-1}\theta^{j-1}(\theta-1)(q^2-1)\right)(q^2-1)q^{2^i\theta^j k}t^{2^i\theta^j k-1}\right)$$

$$= \sum_{i=2}^{\nu_0}\sum_{j=1}^{\nu_1}\left(\left(-2^{i-1}\theta^{j-1}(\theta-1)(q^2-1)\right)\frac{q^{2^i\theta^j}t^{2^i\theta^j-1}}{1-q^{2^i\theta^j}t^{2^i\theta^j}}\right). \tag{14}$$

Then using (3) and combining (4), (11), (12), (13) and (14) we obtain

$$
\begin{aligned}
\frac{L'_\chi(t)}{L_\chi(t)} =\ & -(q^2-1)\frac{q^\theta t^{\theta-1}}{1-q^{2\theta}t^{2\theta}} - (q^2-1)\frac{q^2 t}{1-q^2 t^2} \\
& + \sum_{i=2}^{\nu_0}\left(-2^{i-1}(q^2-1)\frac{q^{2^i}t^{2^i-1}}{1-q^{2^i}t^{2^i}}\right), \\
& + \sum_{j=1}^{\nu_1}\left(-2\theta^{j-1}(\theta-1)(q^2-1)\frac{q^{2\theta^j}t^{2\theta^j-1}}{1-q^{2\theta^j}t^{2\theta^j}}\right), \\
& + \sum_{i=2}^{\nu_0}\sum_{j=1}^{\nu_1}\left((-2^{i-1}\theta^{j-1}(\theta-1)(q^2-1))\frac{q^{2^i\theta^j}t^{2^i\theta^j-1}}{1-q^{2^i\theta^j}t^{2^i\theta^j}}\right).
\end{aligned}
$$

Then by taking the integral of both sides and simplifying the equation we obtain the desired result, which completes the proof.

If we restrict $q+1=2^{\nu_0}$ for some positive integer ν_0 we immediately obtain the following result. Note that there exist infinitely many values ν_0 such that $q+1=2^{\nu_0}$ for a prime power q if and only if Mersenne prime conjecture holds (see, for example, [2, Chapter IX, Lemma 2.7]).

Corollary 1. *Let χ be the curve $y^{q^2} - y = x^{q+1}$ over $\mathbb{F}_{q^2}$ and assume that $q+1=2^{\nu_0}$ for some positive integer ν_0. Then the L-polynomial $L_\chi(t)$ becomes*

$$
\prod_{i=1}^{\nu_0}\left(1-q^{2^i}t^{2^i}\right)^{\frac{q^2-1}{2}}.
$$

Using the similar methods as in the proof of Theorem 3 we obtain the following result. We would like to stress that L-polynomial below depends on the characteristic of $\mathbb{F}_q$.

Proposition 1. *Let χ be the curve $y^q - y = x^{q^4+1} + \alpha$ over $\mathbb{F}_{q^6}$. Assume that $A = \mathrm{Tr}_{\mathbb{F}_{q^6}/\mathbb{F}_q}(\alpha) \neq 0$ and put $p = char(\mathbb{F}_q)$. Then the L-polynomial $L_\chi(t)$ becomes*

$$
L_\chi(t) = \begin{cases}
\dfrac{\left(1-q^{3p}t^p\right)^{\frac{q}{p}}\left(1-q^{24p}t^{8p}\right)^{\frac{q(q^4-1)}{8p}}}{\left(1-q^3 t\right)\left(1-q^{24}t^8\right)^{\frac{q^4-1}{8}}} & \text{if } q \equiv 1 \mod 4, \\[2em]
\dfrac{\left(1+q^{3p}t^p\right)^{\frac{q}{p}}\left(1-q^{24p}t^{8p}\right)^{\frac{q(q^4-1)}{8p}}}{\left(1+q^3 t\right)\left(1-q^{24}t^8\right)^{\frac{q^4-1}{8}}} & \text{if } q \equiv 3 \mod 4.
\end{cases}
$$

4   Appendix

Here we recall some of the results obtained in [8] for completeness. Let p be an odd prime. For positive integers e and m, let $q = p^e$. Let h be a nonnegative integer and $\alpha, \gamma \in \mathbb{F}_{q^m}$ with $\gamma \neq 0$. Recall that $N(m,n)$ denote the cardinality

$$
N(m,n) = \left|\left\{x \in \mathbb{F}_{q^m} \mid \mathrm{Tr}_{\mathbb{F}_{q^m}/\mathbb{F}_{q^n}}\left(\gamma x^{q^h+1} - \alpha\right) = 0\right\}\right|.
$$

We define nonnegative integers s, t and positive integers r, m_1, h_1 as follows:

$$m = 2^s r m_1,$$
$$h = 2^t r h_1,$$

where $\gcd(m_1, h_1) = \gcd(2, r m_1 h_1) = 1$. Furthermore, let u be the nonnegative integer and ρ, n_1, m_2 be the positive integers so that

$$n = 2^u \rho n_1 \quad \text{and} \quad m_1 = n_1 m_2$$

such that $\gcd(2, \rho n_1) = 1$, $\rho | r$ and $n_1 | m_1$. Finally let

$$A = \mathrm{Tr}_{\mathbb{F}_{q^m}/\mathbb{F}_{q^n}} (\alpha).$$

Note that $u \leq s$ as $n | m$.

Theorem 4. *Assume that $s \leq t$. Let η and η' denote the quadratic characters of $\mathbb{F}_q$ and $\mathbb{F}_{q^m}$, respectively. We have*

– *If m/n is even and $A = 0$, then*

$$N(m, n) = \begin{cases} q^{m-n} - (q^n - 1) \, q^{m/2-n} & \text{if } \eta\left((-1)^{m/2}\right) \eta'(\gamma) = 1, \\ q^{m-n} + (q^n - 1) \, q^{m/2-n} & \text{if } \eta\left((-1)^{m/2}\right) \eta'(\gamma) = -1. \end{cases}$$

– *If m/n is even and $A \neq 0$, then*

$$N(m, n) = \begin{cases} q^{m-n} + q^{m/2-n} & \text{if } \eta\left((-1)^{m/2}\right) \eta'(\gamma) = 1, \\ q^{m-n} - q^{m/2-n} & \text{if } \eta\left((-1)^{m/2}\right) \eta'(\gamma) = -1. \end{cases}$$

– *If m/n is odd and $A = 0$, then*

$$N(m, n) = q^{m-n}.$$

– *If m/n is odd, $A \neq 0$ and n is even, then*

$$N(m, n) = \begin{cases} q^{m-n} + q^{(m-n)/2} & \text{if } (u_1, u_2) \in \{(1, 1), (-1, -1)\}, \\ q^{m-n} - q^{(m-n)/2} & \text{if } (u_1, u_2) \in \{(1, -1), (-1, 1)\}, \end{cases}$$

where u_1 and u_2 are the integers in the set $\{-1, 1\}$ given by

$$u_1 = \eta\left((-1)^{m/2}\right) \eta'(\gamma) \text{ and } u_2 = \eta\left((-1)^{n/2}\right) \eta'(A).$$

– *If m/n is odd, $A \neq 0$ and n is odd, then*

$$N(m, n) = \begin{cases} q^{m-n} + q^{(m-n)/2} & \text{if } (u_1, u_2) \in \{(1, 1), (-1, -1)\}, \\ q^{m-n} - q^{(m-n)/2} & \text{if } (u_1, u_2) \in \{(1, -1), (-1, 1)\}, \end{cases}$$

where u_1 and u_2 are the integers in the set $\{-1, 1\}$ given by

$$u_1 = \eta\left((-1)^{(m-1)/2}\right) \eta'(\gamma) \text{ and } u_2 = \eta\left((-1)^{(n-1)/2}\right) \eta'(A).$$

Theorem 5. *Assume that* $s \geq t + 1$ *and* $u \leq t$. *Let* ω *be a generator of the multiplicative group* $\mathbb{F}_{q^m} \setminus \{0\}$ *and let* a *be the integer with* $0 \leq a < q^m - 1$ *such that* $\gamma = \omega^a$.
We have

- *Case* $s = t + 1$: *Put* $q_1 = q^{2^t}r$.
  If $a \not\equiv m_1 \frac{q_1+1}{2} \mod (q_1 + 1)$, *then*

$$N(m,n) = \begin{cases} q^{m-n} + q^{m/2-n} & \text{if } A \neq 0, \\ q^{m-n} - (q^n - 1)q^{m/2-n} & \text{if } A = 0. \end{cases}$$

  If $a \equiv m_1 \frac{q_1+1}{2} \mod (q_1 + 1)$, *then for* $k = 2^{t+1}r$ *we have that*

$$N(m,n) = \begin{cases} q^{m-n} - q^{(m+k)/2-n} & \text{if } A \neq 0, \\ q^{m-n} + (q^n - 1)q^{(m+k)/2-n} & \text{if } A = 0. \end{cases}$$

- *Case* $s \geq t + 2$: *Put* $q_1 = q^{2^t}r$.
  If $a \not\equiv 0 \mod (q_1 + 1)$, *then*

$$N(m,n) = \begin{cases} q^{m-n} - q^{m/2-n} & \text{if } A \neq 0, \\ q^{m-n} + (q^n - 1)q^{m/2-n} & \text{if } A = 0. \end{cases}$$

  If $a \equiv 0 \mod (q_1 + 1)$, *then for* $k = 2^{t+1}r$ *we have that*

$$N(m,n) = \begin{cases} q^{m-n} + q^{(m+k)/2-n} & \text{if } A \neq 0, \\ q^{m-n} - (q^n - 1)q^{(m+k)/2-n} & \text{if } A = 0. \end{cases}$$

Theorem 6. *Assume that* $t + 1 \leq u \leq s$ *and* $A = 0$. *Let* ω *be a generator of the multiplicative group* $\mathbb{F}_{q^m} \setminus \{0\}$ *and let* a *be the integer with* $0 \leq a < q^m - 1$ *such that* $\gamma = \omega^a$.
We have

- *Case* $s = t + 1$: *Put* $B_1 = \gcd\left(m_2, q^{2^t \rho} + 1\right)$.
  If $a \equiv n_1 m_2 \frac{q^{2^t r}+1}{2} \mod \left(\frac{q^{2^t r}+1}{q^{2^t \rho}+1} B_1\right)$, *then*

$$N(m,n) = q^{m-n} - (q^n - 1)q^{m/2-n} + B_1 \frac{q^n - 1}{q^{2^t \rho} + 1}\left(q^{m/2+2^t r-n} + q^{m/2-n}\right).$$

  If $a \not\equiv n_1 m_2 \frac{q^{2^t r}+1}{2} \mod \left(\frac{q^{2^t r}+1}{q^{2^t \rho}+1} B_1\right)$, *then*

$$N(m,n) = q^{m-n} - (q^n - 1)q^{m/2-n}.$$

- *Case* $s \geq t + 2$: *Put* $B_1 = \gcd\left(2^{s-u} m_2, q^{2^t \rho} + 1\right)$.
  If $a \equiv 0 \mod \left(\frac{q^{2^t r}+1}{q^{2^t \rho}+1} B_1\right)$, *then*

$$N(m,n) = q^{m-n} + (q^n - 1)q^{m/2-n} - B_1 \frac{q^n - 1}{q^{2^t \rho} + 1}\left(q^{m/2+2^t r-n} + q^{m/2-n}\right).$$

If $a \not\equiv 0 \mod \left(\frac{q^{2^t r}+1}{q^{2^t \rho}+1} B_1 \right)$, then

$$N(m,n) = q^{m-n} + (q^n - 1)q^{m/2-n}.$$

Finally we note that the only remaining case in the above theorems is $t + 1 \leq u \leq s$ and $A \neq 0$. Together with A. Coşgun we determined $N(\chi)$ also in this case [1]. We note that this case is more complicated.

Acknowledgment. The first author was partially supported by TÜBİTAK under Grant No. TBAG-112T011.

References

1. Coşgun, A., Özbudak, F., Saygı, Z.: Rational points of some algebraic curves over finite fields, in preparation
2. Huppert, B., Blackburn, N.: Finite Groups II. Springer-Verlag, Berlin, Heidelberg, New York (1982)
3. Kedlaya, K.S.: Counting points on hyperelliptic curves using Monsky-Washnitzer cohomology. J. Ramanujan Math. Soc. **16**, 323–338 (2001)
4. Lauder, A.G.B., Wan, D.: Computing zeta functions of Artin-Schreier curves over finite fields. Lond. Math. Soc. JCM **5**, 34–55 (2002)
5. Manin, Y.I.: The theory of commutative formal groups over fields of finite characteristic. Uspekhi Mat. Nauk **18:6**(114), 3–90 (1963)
6. Niederreiter, H., Xing, C.: Rational Points on Curves over Finite Fields: Theory and Applications. Cambridge University Press, Cambridge (2001)
7. Niederreiter, H., Xing, C.: Algebraic Geometry in Coding Theory and Cryptography. Princeton University Press, Princeton (2009)
8. Özbudak, F., Saygı, Z.: Rational points of the curve $y^{q^n} - y = \gamma x^{q^h+1} - \alpha$ over $_{q^m}$. In: Larcher, G., Pillichshammer, F., Winterhof, A., Xing, C.P. (eds.) Applied Algebra and Number Theory. Cambridge Univesity Press, Cambridge (2014)
9. Rück, H.-G.: A note on elliptic curves over finite fields. Math. Comp. **49**, 301–304 (1987)
10. Rück, H.-G., Stichtenoth, H.: A characterization of Hermitian function fields over finite fields. J. Reine Angew. Math. **457**, 185–188 (1994)
11. Stichtenoth, H.: Die Hasse-Witt-Invariante eines Kongruenzfunktionenkorpers. Arch. Math. **33**, 357–360 (1979)
12. Stichtenoth, H.: Algebraic Function Fields and Codes. Springer-Verlag, Berlin (2009)
13. Tate, J.: Endomorphisms of abelian varieties over finite fields. Invent. Math. **2**(2), 134–144 (1966)
14. Tsfasman, M.A., Vladut, S.G., Nogin, D.: Algebraic Geometric Codes: Basic Notions. American Mathematical Society, Providence (2007)
15. Voloch, J.F.: A note on elliptic curves over finite fields. Bull. Soc. Math. France **116**, 455–458 (1989)
16. Waterhouse, W.: Abelian varieties over finite fields. Ann. Sci. Ecole Norm. **2**(4), 521–560 (1969)

Coding Theory and Code-Based Cryptography

Efficient Software Implementations
of Code-Based Hash Functions
and Stream-Ciphers

Pierre-Louis Cayrel[1], Mohammed Meziani[2], Ousmane Ndiaye[3]([⊠]),
and Quentin Santos[4]

[1] Laboratoire Hubert Curien, UMR CNRS 5516, Bâtiment F 18 rue du professeur
Benoît Lauras, 42000 Saint-etienne, France
pierre.louis.cayrel@univ-st-etienne.fr
[2] CASED-Center for Advanced Security Research Darmstadt,
Mornewegstrasse, 64293 Darmstadt, Germany
mohammed.meziani@cased.de
[3] Université Cheikh Anta Diop de Dakar, FST, DMI, LACGAA, Dakar, Senegal
ousmane3.ndiaye@ucad.edu.sn
[4] École Normale Supérieure de Lyon, 46, Allée d'Italie, 69007 Lyon, France
quentin.santos@ens-lyon.fr

Abstract. In this work, we present a survey on software implementations of two families of cryptographic primitives based on the syndrome
decoding problem: hash functions and stream ciphers. We have studied different algorithms, namely, FSB, SFSB, RFSB, SYND, 2SC and
XSYND, and tried to improve their performances as software implementations which are done in C language by Using XMM registers from
Streaming SIMD Extensions (SSE). We provide a fair comparison of the
implementations of those primitives in the same platform and also give
links to the codes we have developed. Although we did not reach the
speed given in the paper in some cases, we managed to beat the results
of the reference implementations when they are available.

Keywords: Code-based cryptography · Software implementations

1 Introduction

Informally, a hash function is defined as a function that compresses an input
of arbitrary length into a string with a fixed length, called hash or digest. If
it satisfies additional requirements, it can be used as a powerful tool in cryptography to secure information infrastructure such as: secure web connections,
encryption key management, virus and malware-scanning, password logins, and

Supported in part by NATO's Public Diplomacy Division in the framework of
"Science for Peace", SPS Project 984520.
Supported by the Pole of Research in Mathematics and their Applications in Infor-
mation Security (PRMAIS).

digital signatures. Over the last years, a growing number of hash functions were proposed, but unfortunately most of the proposals used in practice have been found to be vulnerable and should not be used. This has called into question the long-term security of later proposals, whose designs are derived from these hash functions like SHA-2 family. As a reaction, the US National Institute of Standards and Technology (NIST) announced the public SHA-3 competition (or the Advanced Hash Standard (AHS)), to develop new cryptographic hash algorithms. Initially, NIST received 64 proposals, but only 5 candidates have been selected to advance to the third (and final) round of the contest. One of the 64 submissions is the Fast Syndrome-Based hash Function (in short FSB) introduced first by Augot, Finiasz, and Sendrier [FSB] in 2003 and improved by Finiasz, Gaborit, and Sendrier in 2007 in [FSB-SHA3]. This hash function is still unbroken up to date and is provably secure, meaning that its security is directly reducible to NP-complete problems from coding theory, that are believed to be hard on average. However, it suffers from the disadvantage of being inefficient compared to the other competing hash functions. This issue is the main reason of being deselected in the second round of the competition. Motivated by its inefficiency, another hash function based on the syndrome decoding problem has been proposed by Meziani et al. in [SFSB] which improves the speed of FSB by using the sponge construction. More recently, Bernstein et al. have proposed an even faster hash-function in [RFSB] which is as fast as 2 of the 5 SHA-3 finalists with less than 14 cycles to hash one byte for 128 bit security level.

In contrast to hash functions, stream ciphers are a very important family in secret key encryption schemes, where each plaintext block is combined, one at a time, with the corresponding block of a pseudo-random sequence (called the keystream) to produce one block of the ciphertext. Typically, the combination is done by means of the bitwise XOR operation. Stream ciphers have to be exceptionally fast and require low computing resources. Therefore, they are used in many applications like GSM, UMTS, RFID, Bluetooth and online encryption of big amounts of data in general.

The theoretically secure constructions of PRNG(Pseudo-Random Number Generator) have, up until now, mostly been focused on methods based on number theoretic assumptions. Despite their simplicity, most of these systems are inefficient thus impractical for many applications. It is therefore desirable to have efficient stream ciphers whose security relies on other assumptions. The first construction, based on the syndrome decoding problem, has been proposed by Fisher and Stern [6]. Gaborit et al. proposed SYND [SYND] which is an improved variant of Fisher-Stern's system. With 2SC [2SC], Meziani et al. proposed a faster stream cipher using the so-called sponge construction. More recently, Meziani et al. [XSYND] proposed the XSYND stream cipher as an improved variant of SYND in terms of performance. Their proposal outperforms all previous code-based constructions of stream ciphers.

Our Contribution: We have implemented each of the recently proposed hash functions and stream ciphers. Some implementations have been improved by some low-level optimization or simpler code. We have run benchmarks on a unique platform for performance comparison.

Organization of the Paper: We first present the hash functions (FSB, SFSB, RFSB) and the stream ciphers (SYND, 2SC, XSYND) we have worked on. Then, we focus on the choice of the parameters for each function for both security and efficiency matters. Considering those functions with theses parameters, we give the results of our benchmarks. Finally, we detail how we obtained such results and our optimizations.

2   Cryptographic Basis

2.1   Cryptographic Primitives

We focus on two main primitives: hashing functions and stream ciphers. The former intends to provide a hard-to-reverse function, which does not require any key and whose output length is fixed. The latter intends to provide an easy-to-reverse function, protected by a key and whose output length may vary.

A hashing function must ensure that the following operation are hard to perform:

- preimage resistance ($f(?) = y$): for a given hash, it is hard to compute a message which result in the same hash value when passed through the hash function;
- second preimage resistance ($f(?) = f(x)$): for a given message, it is hard to find another one which has the same hash;
- collision resistance ($f(?) = f(?)$): it is hard to find two messages with the same hash.

2.2   Splitting and Padding

Almost all hashing functions and stream ciphers work on message blocks instead of taking the message as a whole. Thus, the message is first split into blocks of a fixed length as shown below:

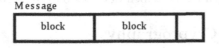

Fig. 1. Message Splitting

In Fig. 1, the message is split and the blocks have the same length maybe except for the last one. As the last block may not be as long as the other ones, some bits are added to complete it:

Message

Padding

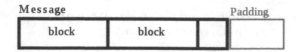

Fig. 2. Message Padding

In Fig. 2, Some bits are added to fill the last block. To avoid trivial collision (adding zeros to the message would give the same hash), the length of the initial message is to be stored in the padding.

Once the message has been split and padded, the primitive handles one block after another. The important point is that the chaining must be designed such that any change in a block changes the behavior of the function all the way to the end.

2.3 PRNG Based Stream-Cipher

Most of the recent stream cipher definitions have the following approach: the input message (plain text) is split into blocks and padded as explained in the previous subsection. The encryption performing itself is done by a very simple operation, like a bitwise XOR. The security is based on the one operand, which is given by the precedent keystream as a session key (bit strings).

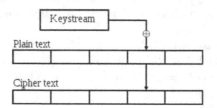

Fig. 3. Stream cipher scheme

In Fig. 3, every message block is XORed to a single keystream. A keystream is a block which takes a cipher key and outputs a deterministic sequence of numbers which are not predictable without the key. In fact, the keystream is produced by a Pseudo (or Deterministic) Random Number Generator (PRNG) which is initialized with a seed which depends on the secret key.

3   Code-Based Cryptography

3.1   Coding Theory

In this paper, we study functions which rely on a well known problem of linear codes. A code is a way to represent information. Simply speaking, it is a set of *words* C. A linear $C(n, k)$ is a vector subspace of dimension k; of a vector space of dimension $n > k$. A such vector space can be seen as a kernel of a linear application represented by a $(n - k) \times n$ matrix H.

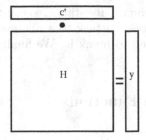

Fig. 4. The syndrome decoding problem

3.2 The Syndrome Decoding (SD) Problem

Let us consider the following equation: $H.c' = y$ where $y \neq 0$. y is called a *syndrome*.

The Syndrome Decoding (SD) problem is, for a given syndrome y, a linear code C given by a matrix H, and a weight t, to compute c' of weight t such that $H.c' = y$.

In Fig. 4, H's columns corresponding to the positions of non zero bits of c' are XORed. The larger t is, the harder the problem is.

3.3 Complexity Reduction

To estimate the security of a cryptographic primitive (hash function or PRNG) based on SD, one first studies the known attacks on SD. Then, one reduces the attacking of the primitive to the attacking of SD and deduces the minimum complexity of an attack. This way, a security level can be given for various parameters.

The point of relying on an NP-hard problem is to ensure a certain level of security. Moreover, unlike RSA, the code-based protocols are still secure against quantum algorithms. This is sometimes referred to as post-quantum cryptography.

3.4 Choice of the Parameters

We will have to choose parameters for the functions, which will increase the speed and the security of the algorithm. The speed is usually given in cycles per byte, *id est* the mean number of cycles during which the processor runs the algorithm to process one byte of input (hash function) or to output one pseudo-random byte (PRNGs). The security is given as the logarithm of the number of operations an attacker would need to break the secrecy: sec = $log_2(n_{\text{ops}})$. A system is currently considered secure if the security is above 128.

Thus, we need to know the minimum number of operations needed to bypass the functions. In order to estimate it, we first study some well-known attacks against SD. Here, the best attacks are the Information Set Decoding (ISD), the Generalized Birthday Attack (GBA), and the linearization attack. Then, for each security point (preimage resistance, second preimage resistance, and collision resistance) we reduce an attack to a SD instance thereby deducing the number of operations required to break it. We finally keep the smaller number as a ceiling for the security.

4   Code-Based Hash Functions

4.1   Merkle-Damgård

We use the Merkle-Damgård [Merkle79, Merkle89, Damgard90] scheme, the first realistic to chain the block hashing from a message of any size.

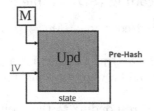

Fig. 5. Merkle-Damgård scheme

In Fig. 5 is represented the black box function (update) of the Merkle-Damgård domain.

The Merkle-Damgård domain extender is a method of building collision-resistant cryptographic hash functions from collision-resistant one-way compression functions. The principle is to split any message into blocks and to process each of them one after another through an update function. The importance of this scheme is that the result (the state) is cycled back to the function to spread any change in the message. The state is initialized with a value "IV" (Initialization Vector) and the final state is called the "pre-hash". The "pre-hash" can then be processed as an "IV" to get new "pre-hash" or the final hash (for instance, through the Whirlpool algorithm).

4.2   FSB

FSB [FSB] (Fast Syndrome Based hash function) was first proposed in 2003 as a replacement of SB [SB] (Syndrome Based hash function). It combines some bits of the state and some bits of the message block to compute an index. Actually, w indexes are computed as such, and the w columns associated with these indexes are XORed together to get the output.

The Fig. 6 represents the FSB update function based on the Merkle-Damgård model (see Fig. 5) and a syndrome computing.

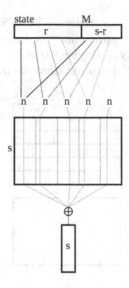

Fig. 6. FSB update function

Here is a sum up of the FSB parameters:

n	number of available columns
w	number of columns to be XORed
s	length of the XORed vectors
r	unmodified part of the state
p	length of the shifted vectors

FSB was proposed as a SHA-3 candidate [FSB-SHA3] but failed on round two, due to its lack of speed. The parameters given for this version are as follows:

Hash length	n	w	r	p	s
160	20×2^{16}	80	640	653	1120
224	28×2^{16}	112	896	907	1568
256	32×2^{16}	128	1024	1061	1792
384	23×2^{16}	184	1472	1483	2392
512	31×2^{16}	248	1984	1987	3224

4.3   SFSB

SFSB [SFSB] (Sponge for Fast Syndrome Based hash function) was proposed in 2011 and is intended to increase the speed of FSB by "merging" the state and the message block, instead of concatenating it. SFSB works quite the same way as FSB except that, instead of appending the message block, it is XORed with the state. The sponge capacity is the length of the part of the state which has not been XORed. Moreover, the computation of the indexes is simplified.

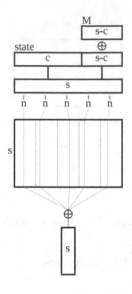

Fig. 7. SFSB update function

The Fig. 7 represents the SFSB update function based on the Merkle-Damgård model (see Fig. 5) and a syndrome computing without concatenating but merging the state and the message block.

Here is a sum up of the SFSB parameters:

n	number of available columns
w	number of columns to be xored
s	length of the xored vectors
c	sponge capacity
p	length of the shifted vectors

The specifications of SFSB give the set of parameters below:

Hash length	n	s	w	c	Nxor	GBA_p	ISD_p	GBA_c	ISB_c
160	12×2^{17}	384	24	240	64.0	130	99	86	91
224	17×2^{17}	544	34	336	88.9	150	144	114	112
256	39×2^{17}	624	39	296	90.5	246	172	129	148

GBA_p and ISD_p stands for the complexity of a preimage attack using GBA or ISD whereas GBA_c and ISD_c are for a collision attack.

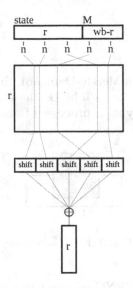

Fig. 8. RFSB update function

4.4 RFSB

RFSB [RFSB] (Really Fast Syndrome Based hash function) was also proposed in 2011 and slightly changed the way the quasi-cyclic matrices are handled. The shift is made during the sum up, depending only on the rank in the sum. It also takes advantage of hardware considerations, greatly improving the efficiency. RFSB selects the XORed columns in a simpler way, which strongly alleviates the strain on the matrix (wording is a bit weird). Theses columns are then shifted and properly XORed. This approach was chosen to simplify the implementations as mush as possible.

The Fig. 8 represents the RFSB update function based on the Merkle-Damgård model (see Fig. 5) on a syndrome computing and on a permutation of the syndrome.

Here is a sum up of the RFSB parameters:

r	length of the XORed vectors
b	bits read at each step
w	number of steps for each block

The specifications of RFSB only take into account one set of parameters, this set is intended to use the hardware registers at their best level of performance

r	b	w
509	8	112

5   PRNGs

5.1   General Scheme

As for the hash functions with Merkle-Damgård, the PRNGs we study have the same general structure. Here, for each block, the state is updated and an output (the XOR mask used to encrypt) is processed from this state.

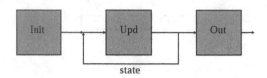

Fig. 9. PRNGs scheme

The Fig. 9 represents the PRNG scheme based on the update function of the Merkle-Damgård model (see Fig. 5).

A PRNG is described as an update function and an output function (and the initialization function).

5.2   SYND

SYND [SYND] was published in 2007 as a code-based PRNG. Here, both the update function and the output function work like they do for FSB. For a given matrix F, we define, f, a function which computes a new state or output from the last state.

SYND uses two functions, f_1 and f_2 defined by the matrixes F_1 and F_2. They are both used in the initialization function.

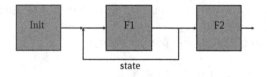

Fig. 10. SYND scheme

The Fig. 10 represents the SYND scheme based on the PRNGs scheme (see Fig. 9) and the FSB update function (see Fig. 6).

The description of the parameters:

k	key length
n	number of available columns
r	length of the XORed vectors

n	k	r
8192	64	128
8192	96	192
8192	128	256
8192	192	384
8192	256	512
8192	512	1024

5.3   2SC

In 2011, 2SC [2SC] (Sponge for Code-based Stream Cipher) proposed the use of the sponge construction on SYND. The update function works the same as it did for SYND (function h_1 defined by a matrix H_1) but the output is computed by truncating the state. A second function, h_2, defined by a matrix, H_2, is used alongside h_1 in the initialization function.

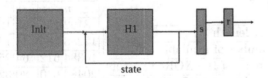

Fig. 11. 2SC scheme

The Fig. 11 represents the 2SC scheme based on the SYND scheme (see Fig. 10) by truncating the state for outputs.

Description of the parameters:

n	number of available columns	n	s	w	c
s	length of the XORed vectors	1572864	384	24	240
c	sponge capacity	2228224	544	34	336
w	number of XORed columns	3801088	928	58	576

5.4   XSYND

XSYND [XSYND] has been published in 2012. It intends to use the RFSB scheme in a PRNG structure. Unlike for SYND, here, the functions are described as they are for RFSB:

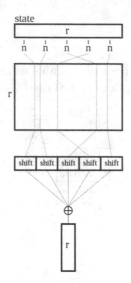

Fig. 12. FSB style function

The Fig. 12 represents the FSB style function used for the XSYND scheme. Parameters:

			n	r	w	k
k	key length		8192	256	32	128
n	number of available columns		12288	384	48	192
r	length of the XORed vectors		16384	512	64	256
w	number of XORed columns		20480	640	80	320
			24576	768	96	384
			28672	896	112	448

6   Optimization Matters

Here, we describe the optimizations we have used to improve the speed of the hash functions and stream ciphers. For a better comprehension of the code, a 'plain' version contains the simplest version of our implementations. We also provide the intermediate values of these functions.

6.1   Cleaning Code

Although we wrote our code from scratch, we referred to the reference implementation for a better understanding of the paper and disambiguation. In most cases, the code was more like a translation of the theoretical description than a description of how the computer should work at a low level. Thus, we avoided (almost) all of the useless operations and moved the choice of the parameters to the C preprocessor. It was a noticeable improvement.

6.2 Classic Registers

The first step in the actual optimization was to make use of all of the registers available on the processor. We changed the code so that the computing would use 8, 32, or 64 bit variables depending on compilation options. In some cases, it accelerated the functions by a factor greater than 2.

6.3 XMM Registers

The XMM extensions introduced eight new 128 bit data registers and SSE2, made their use very easy. SSE2 is now present on almost all processors, allowing a wide public use. Every register use packets which contain floating numbers and integers, and supports several instructions. Moreover, the 64 bit processors provide eight more such registers. These sixteen registers are named XMM0 to XMM15.

Generally by using low-level programming (ASM), SSE processors convert data into a vector or SIMD(Single Instruction Multiple Data) units and process in parallel programming. Its approach is similar to integrated circuit such as FPGA:

1. Load up sixteen data points at once into SIMD registers
2. The taken Operation being applied to the data will happen to all sixteen values at the same time
3. Set the result in memory

Thereby, the principle of vectorization allows us to speed up programs.

We first adapted our implementations to use the first eight registers at their best but using more of them was useful: for some parameters FSB state has a length which does not fit in the registers and RFSB needs a ninth temporary register. In those cases, we had to use memory to fill the lack of registers.

Although these first implementations are still available as fsb32.c, sfsb32.c, and rfsb32.c for compatibility, we then allowed ourselves to use the eight more registers. The best performance improvement concerned SFSB-384 as our previous implementation of it did not use XMM registers at all.

In comparison to the 64 bit implementations, the improvement was still important.

6.4 Loop Unrolling

The update functions are each made of a loop which reads the inputs to compute indexes. Nevertheless, since the input length is fixed, so is the loop. Now, a fixed length loop uses classical loop operations even though it just has to repeat the same code a fixed number of times. Instead of a classical loop with conditional

Function	speed in cpb (art.)	speed in cpb (ref.)	speed in cpb (our)	ratio our / art.	ratio our / ref.	Security
FSB-160	257	395.12	110.21	0.43	0.28	100
FSB-224	297	482.51	131.50	0.44	0.27	135
FSB-256	324	528.39	137.41	0.42	0.26	153
FSB-384	423	652.20	203.99	0.48	0.31	215
FSB-512	507	820.87	264.33	0.52	0.32	285
SFSB-160	160	286.20	79.82	0.50	0.28	86
SFSB-224	201	422.91	99.92	0.50	0.24	114
SFSB-384	183	604.19	172.65	0.94	0.29	129
RFSB-509	13.62	22.82	17.26	1.27	0.76	142

Fig. 13. Performance of hash functions

jumps, condition tests, counter increments and memory access for the counter storing, one can unroll the loop by directly repeating the code in the executable. This way, we trade some space (code segment) for some time.

A partial unrolling does not remove the loop but simplifies some iterations. For instance, the first iteration is removed from the loop to be optimized alone.

We run benchmarks to find the best choice for each function: partial unrolling, manual unrolling (see fsb32.c and outXorVi.c), or automatic unrolling (compiler option *-funroll-loops*).

6.5 Data Representation

For comparison purposes we make implementations with the same parameters as original articles (cf. Sect. 4) and run the reference implementation on the same platform: *AMD Phenom II X2 550 processor* (64 bits) operated by a *Debian GNU/Linux (3.2.0-2-amd64) wheezy/sid*.

Due to XMM registers, *byte* is used as the unit of information on all implementations, and memory allocation is done by a contiguous manner.

As our strategy is to use SSE, we found it necessary to adapt the data storage in memory to exploit the XMM registers by taking advantage of parallelism.

So for schemes based on a matrix representation (FSB, SFSB), we choose to split by sub-matrices to make easy the selection of a single column in each sub-matrix during process. For those that are based on the polynomial rings (XSYND, RFSB), we use the polynomial(vector) representation.

For 2SC and SYND we use two matrices to represent both linear mappings used.

7 Results

7.1 Benchmarks

The following benchmarks have been run on a *AMD Phenom II X2 550 processor* (64 bits) operated by a *Debian GNU/Linux (3.2.0-2-amd64) wheezy/sid* on a

Function	speed in cpb (art.)	speed in cpb (ref.)	speed in cpb (our)	ratio our / art.	ratio our / ref	Security
SYND-64	22	-	23.88	1.09	-	71
SYND-96	46	-	29.91	0.65	-	102
SYND-128	27	108.08	30.27	1.12	0.28	131
SYND-192	47	-	53.80	1.14	-	191
SYND-256	53	280.76	41.50	0.78	0.15	252
SYND-512	83	430.36	149.94	1.81	0.35	493
2SC-144	37	298.97	25.18	0.68	0.08	80
2SC-208	47	387.55	33.22	0.71	0.09	170
2SC-352	72	605.61	80.05	1.11	0.13	366
XSYND-509	-	55.36	15.25	-	0.27	142

Fig. 14. Performance of PNRGs and Stream Ciphers

random file of 100MiB generated from */dev/urandom*. The execution time is the user time measured using the Unix tool, *time*.

The tables below give information about the speed of these functions. The first three columns give the speed (in mean number of cycles of processor required to process one byte) given in the article, the one we got with the reference implementation (when applicable), and the one of ours. The security level is given by its equivalent in bit-length of the Work Factor ($log_2(WF)$).

The formula for cpb is: execution time (s) × processor frequency (Hz) / test file size (bytes).

We found reference implementations for FSB[1], SFSB and 2SC[2] and for RFSB[3].

7.2   Discussions

Implementations Comparison. This comparison based on the execution speed of the algorithms, is measured by the ratio between the number of cpb our implementation and those of the reference implementation, because the two sources are executed on the same platform.

Hash functions: Except for RFSB, we actually improved the speed of all the hash functions. The main explanation is that the previous implementations were not optimized. Concerning RFSB, we could not reach the announced speed with the implementation we used as a reference, but our implementation is faster than the reference implementation. Most likely, the implementation the authors of the article used benefited more machine-specific optimizations performed by the compiler or used optimization techniques which are not described in the paper.

---

[1] https://www.rocq.inria.fr/secret/CBCrypto/index.php?pg=fsb.
[2] http://www.cayrel.net/research/code-based-cryptography/code-based-crypto systems/article/implementation-of-code-based-hash.
[3] http://bench.cr.yp.to/supercop.html.

For FSB, the best (lowest) ratio equals to 0.26 is obtained with FSB-256 and the worst ratio equals to 0.32 with FSB-512.

For SFSB, the best ratio equals to 0.24 with SFSB-224 and a worst ratio equals to 0.29 with SFSB-384.

For RFSB-509, we improved the reference implementation with a ratio equals to 0.76, but we don't get better speed than the one which is on the original article [RFSB] (ratio =1.27).

PRNGs: Here, the results are more balanced. We still have better results than the reference implementations we found but we could not always get better results than the article. In two cases (SYND-512 and 2SC-352), the speed is far worse than for the others. Although our implementations work the same for all the parameters and we expected the speed progression to be linear, the speed does not seem to be regular at all.

For SYND we get the best ratio equals to 0.15 for SYND-256 and a worst ratio equals to 0.35 for SYND-512.

For 2SC we get the best ratio equals to 0.08 for 2SC-144 and a worst ratio equals to 0.13 for 2SC-352.

For XSYND-509, we improved the reference implementation with a ratio equals to 0.27.

Functions Comparison

Hash functions: The fastest hash function among these is RFSB. Its security level is between the ones of FSB-224 and FSB-256. We can thereby estimate that RFSB improves the speed by a factor of **7**. The results of SFSB are more balanced here. For a security level around 100, we do get a ten percent improvement but our SFSB-384 seems to run far slower than for the others parameters. The parameter might forbid some optimization made by our compiler.

PRNGs: Finally, we can compare SYND and 2SC. Since the security levels are not close, we will compare the security levels for close speeds. Around 30 cpb, 2SC provides a good improvement on security. Thus, the sponge construction is clearly shown to be more efficient.

8   Conclusion

We have studied FSB, SFSB, RFSB, SYND, 2SC and XSYND and tried to improve their performances as software implementations. Although we did not reach the speed given in the paper in some cases, we managed to beat the results of the reference implementations when they were available.

The Syndrome Decoding problem lead to a lot of ideas of hash functions and PRNGs. Though, it is not always easy to compare the speed of each because source codes are not always available to be used on the same platform or they are not generally optimized.

In future work, in order to improve these implementation, it would be interesting to bring this study on others platforms such as the ARMv8 architecture or processors that support vector instructions (AVX, AVX2).

References

[2SC] Meziani, M., Cayrel, P.-L., El Yousfi Alaoui, S.M.: 2SC: an efficient code-based stream cipher. In: Kim, T.-H., Adeli, H., John Robles, R., Balitanas, M.O. (eds.) ISA 2011. CCIS, vol. 200, pp. 111–122. Springer, Heidelberg (2011)

[Damgard90] Damgaard, I.B.: A design principle for hash functions. In: Brassard, G. (ed.) Advances in Cryptology (CRYPTO 1989). LNCS, vol. 435, pp. 416–427. Springer, Heidelberg (1990)

[FSB] Augot, D., Finiasz, M., Sendrier, N.: A fast provably secure cryptographic hash function. IACR Cryptology ePrint Archive 2003:230 (2003)

[FSB-SHA3] Finiasz, M., Gaborit, P., Sendrier, N., Manuel, S.: Sha-3 proposal: Fsb. Proposal of a hash function for the NIST SHA-3 competition, October 2008

[Merkle79] Merkle, R.C.: Secrecy, authentication, and public key systems. PhD thesis, Stanford University (1979)

[Merkle89] Merkle, R.C.: A certified digital signature. In: Brassard, G. (ed.) Advances in Cryptology (CRYPTO 1989). LNCS, vol. 435, pp. 218–238. Springer, Heidelberg (1990)

[RFSB] Bernstein, D.J., Lange, T., Peters, C., Schwabe, P.: Really fast syndrome-based hashing. IACR Cryptology ePrint Archive 2011:74 (2011)

[SB] Fischer, J.-B., Stern, J.: An efficient pseudo-random generator provably as secure as syndrome decoding. In: Maurer, U. (ed.) Advances in Cryptology–EUROCRYPT 1996. Lecture Notes in Computer Science, vol. 1070, pp. 245–255. Springer, Heidelberg (1996)

[SFSB] Meziani, M., Dagdelen, Ö., Cayrel, P.-L., El Yousfi Alaoui, S.M.: S-FSB: an improved variant of the FSB hash family. In: Kim, T.-H., Adeli, H., Robles, R.J., Balitanas, M.O. (eds.) ISA 2011. CCIS, vol. 200, pp. 132–145. Springer, Heidelberg (2011)

[SYND] Gaborit, P., Lauradoux, C., Sendrier, N.: Synd: a very fast code-based cipher stream with a security reduction. In: Proceedings of the 2007 IEEE International Symposium on Information Theory - ISIT 2007, pp. 186–190. Nice, France, June 2007

[XSYND] Meziani, M., Hoffmann, G., Cayrel, P.-L.: Improving the performance of the SYND stream cipher. In: Mitrokotsa, A., Vaudenay, S. (eds.) Progress in Cryptology - AFRICACRYPT 2012. Lecture Notes in Computer Science, vol. 7374, pp. 99–116. Springer, Heidelberg (2012)

Quadratic Residue Codes over $\mathbb{F}_p + v\mathbb{F}_p + v^2\mathbb{F}_p$

Yan Liu[1], Minjia Shi[1]([✉]), and Patrick Solé[2]

[1] School of Mathematical Sciences, Anhui University, Hefei, China
smjwcl.good@163.com
[2] CNRS/LTCI Telecom Paris Tech, 46 rue Barrault, 75013 Paris, France

Abstract. This article studies quadratic residue codes of prime length q over the ring $R = \mathbb{F}_p + v\mathbb{F}_p + v^2\mathbb{F}_p$, where p, q are distinct odd primes. After studying the structure of cyclic codes of length n over R, quadratic residue codes over R are defined by their generating idempotents and their extension codes are discussed. Examples of codes and idempotents for small values of p and q are given. As a by-product almost MDS codes over $\mathbb{F}_7$ and $\mathbb{F}_{13}$ are constructed.

Keywords: Cyclic codes · Quadratic residue codes · Generating idempotents · Dual codes

MSC (2010): Primary 94B15; Secondary 11A15.

1 Introduction

Cyclic codes form an important subclass of linear block codes, studied from the fifties onward. Their clean algebraic structure as ideals of a quotient ring of a polynomial ring makes for an easy encoding. Quadratic residue codes are a special kind of cyclic codes of prime length introduced to construct self-dual codes by adding an overall parity-check. Quadratic residue codes over finite fields have been studied extensively by Assmus and Mattson in a series of research reports [1]. Since then, coding theorists have studied quadratic residue codes and their properties over rings that are not fields. Cyclic codes and quadratic residue codes over $\mathbb{Z}_4$ were studied in [2]. Tan X investigated a family of quadratic residue codes over $\mathbb{Z}_{2^m}$ in [3]. The above rings are local. In [4], Kaya A et al. studied quadratic residue codes over $\mathbb{F}_p + v\mathbb{F}_p$ and their Gray images. In [5], Zhang T et al. used another way to prove that some properties of quadratic residue codes over $\mathbb{F}_p + v\mathbb{F}_p$. Cyclic codes and the weight enumerators of linear codes over $\mathbb{F}_2 + v\mathbb{F}_2 + v^2\mathbb{F}_2$ were given in [6]. It is worth mentioning that Kaya A et al. have done great contribution to new extremal and optimal binary self-dual codes from quadratic residue codes over $\mathbb{F}_2 + u\mathbb{F}_2 + u^2\mathbb{F}_2$ in [7].

This research is supported by NNSF of China (61202068), Talents youth Fund of Anhui Province Universities (2012SQRL020ZD). Technology Foundation for Selected Overseas Chinese Scholar, Ministry of Personnel of China (05015133) and the Project of Graduate Academic Innovation of Anhui University (N0. yfc100005).

© Springer International Publishing Switzerland 2015
Ç. Koç et al. (Eds.): WAIFI 2014, LNCS 9061, pp. 204–211, 2015.
DOI: 10.1007/978-3-319-16277-5_12

Following the above trend, this paper is devoted to studying quadratic residue codes over the non-local ring $\mathbb{F}_p + v\mathbb{F}_p + v^2\mathbb{F}_p$, where p is an odd prime, thus generalizing [7]. First, we introduce the structure of cyclic codes over $\mathbb{F}_p + v\mathbb{F}_p + v^2\mathbb{F}_p$, and prove that there is only one idempotent generator for each code. Next, we define the quadratic residue codes over R, and derive a closed form expression for their idempotents.

The material is organized as follows. The next section contains the basics of codes over rings that we need for further notice. Section 3 derives the structure of cyclic codes over R. Section 4 introduces quadratic residue codes, and Sect. 5 studies their extensions. Section 6 contains a numerical example and Sect. 7 draws conclusions and suggests some open problems.

2   Preliminary Results

Throughout, we let R denote the commutative ring $\mathbb{F}_p + v\mathbb{F}_p + v^2\mathbb{F}_p$, where $v^3 = v$, where p is an odd prime. R is a characteristic p ring of size p^3. Denote by η_1, η_2, η_3 respectively the following elements of R.

$$\eta_1 = 2^{-1}v + 2^{-1}v^2,\ \eta_2 = -2^{-1}v + 2^{-1}v^2,\ \eta_3 = 1 - v^2.$$

A direct calculation shows that $\eta_i^2 = \eta_i$, $\eta_i\eta_j = 0$, $\sum_{i=1}^{3}\eta_i = 1$, where $i, j = 1$, 2, 3 and $i \neq j$. The decomposition theorem of ring theory tells us that $R = \eta_1 R \oplus \eta_2 R \oplus \eta_3 R$. A code C of length n over R is an R-submodule of R^n. An element of C is called a codeword of C. A generator matrix of C is a matrix whose rows generate C. The Hamming weight of a codeword is the number of non-zero components. Let $x = (x_1, x_2, \cdots, x_n)$ and $y = (y_1, y_2, \cdots, y_n)$ be two elements of R^n. The Euclidean inner product is given as $(x, y) = \sum_{i=1}^{n} x_i y_i$. The dual code $C^\perp$ of C with respect to the Euclidean inner product is defined as $C^\perp = \{x \in R^n | (x, y) = 0, \text{ for } \forall\ y \in R^n\}$. C is self-dual if $C = C^\perp$, C is self-orthogonal if $C \subseteq C^\perp$.

In the sequel we let $R_q := R[x]/(x^q - 1)$. A polynomial $f(x)$ is abbreviated as f if there is no confusion. An idempotent is an element $e \in R_q$ such that $e^2 = e$. We characterize all idempotents in R_q in the following lemma.

Lemma 2.1. $\eta_1 f_1 + \eta_2 f_2 + \eta_3 f_3$ is an idempotent in R_q if and only if f_i are idempotents in $\mathbb{F}_p[x]/(x^q - 1)$, where $i = 1$, 2, 3.

Proof. Let $g = \eta_1 f_1 + \eta_2 f_2 + \eta_3 f_3$ be an idempotent in R_q, then we have $[\eta_1 f_1 + \eta_2 f_2 + \eta_3 f_3]^2 = \eta_1 f_1^2 + \eta_2 f_2^2 + \eta_3 f_3^2 = \eta_1 f_1 + \eta_2 f_2 + \eta_3 f_3$, which implies $f_i^2 = f_i$, where $i = 1$, 2, 3.

Conversely, if f_i are idempotents in $\mathbb{F}_p[x]/(x^q - 1)$, then $\eta_1 f_1 + \eta_2 f_2 + \eta_3 f_3$ is an idempotent in R_q, since $[\eta_1 f_1 + \eta_2 f_2 + \eta_3 f_3]^2 = \eta_1^2 f_1^2 + \eta_2^2 f_2^2 + \eta_3^2 f_3^2 = \eta_1 f_1 + \eta_2 f_2 + \eta_3 f_3$.

For all $a \in \mathbb{F}_p^*$, the map $\mu_a : \mathbb{F}_p \to \mathbb{F}_p$ is defined by $\mu_a(i) = ai \pmod{p}$ and it acts on polynomials as $\mu_a(\sum_i h_i x^i) = \sum_i h_i x^{\mu_a(i)}$, h_i denote coefficients. It is easily observed that $\mu_a(fg) = \mu_a(f)\mu_a(g)$ for polynomials f and g in R_q.

Let S be a commutative ring with identity, according to the results in [2], then we have the following results.

Theorem 2.2. Let C and D be cyclic codes of length n over S generated by the idempotents f_1, f_2 in $S[x]/(x^n - 1)$, then $C \cap D$ and $C + D$ are generated by the idempotents $f_1 f_2$ and $f_1 + f_2 - f_1 f_2$, respectively.

Theorem 2.3. Let C be a cyclic code of length n over S generated by the idempotent f in $S[x]/(x^n - 1)$, then its dual $C^{\perp}$ is generated by the idempotent $1 - f(x^{-1})$.

Definition 2.4. For any $r' \in R$, r' can be uniquely expressed as $r' = \eta_1 a' + \eta_2 b' + \eta_3 c'$, where $a', b', c' \in \mathbb{F}_p$, the Gray map Φ from R to $\mathbb{F}_p^3$ is defined as $\Phi(\eta_1 a' + \eta_2 b' + \eta_3 c') = (a' + c', a' + b', b' + c')$, which can be generalized to R^n to $\mathbb{F}_p^{3n}$ naturally.

3   Cyclic Codes over $\mathbb{F}_p + v\mathbb{F}_p + v^2\mathbb{F}_p$

In this section, we recall some results on cyclic codes over R, in particular that there is only one idempotent $e \in R_q$ such that $C = (e)$.

If A, B are codes over R, we write $A \oplus B$ to denote the code $\{a+b|a \in A, b \in B\}$. Let C be a linear code of length n over R, we define $C_1 = \{a \in \mathbb{F}_p^n | \exists b, c \in \mathbb{F}_p^n | \eta_1 a + \eta_2 b + \eta_3 c \in C\}$, $C_2 = \{b \in \mathbb{F}_p^n | \exists a, c \in \mathbb{F}_p^n | \eta_1 a + \eta_2 b + \eta_3 c \in C\}$, $C_3 = \{c \in \mathbb{F}_p^n | \exists a, b \in \mathbb{F}_p^n | \eta_1 a + \eta_2 b + \eta_3 c \in C\}$, then $C_i(i = 1, 2, 3)$ are linear codes of length n over $\mathbb{F}_p$, $C = \eta_1 C_1 \oplus \eta_2 C_2 \oplus \eta_3 C_3$ and $|C| = |C_1||C_2||C_3|$.

The following two theorems can be found in [8], we give the proof for copmpleteness.

Theorem 3.1. [8] Let $C = \eta_1 C_1 \oplus \eta_2 C_2 \oplus \eta_3 C_3$ be a cyclic code of length n over R, then $C = (\eta_1 f_1', \eta_2 f_2', \eta_3 f_3')$, where $C_i = (f_i')$, $f_i' \in \mathbb{F}_p[x](i = 1, 2, 3)$, $f_i'|x^n - 1$, and $|C| = p^{3n - \sum_{i=1}^{3} \deg(f_i')}$.

Proof. First, we prove that if $C = \eta_1 C_1 \oplus \eta_2 C_2 \oplus \eta_3 C_3$ is a cyclic code over R if and only if $C_i(i = 1, 2, 3)$ are cyclic codes over $\mathbb{F}_p$. For any $(\alpha_0, \alpha_1, \ldots, \alpha_{n-1}) \in C$, where $\alpha_i = \eta_1 a_i + \eta_2 b_i + \eta_3 c_i, i = 0, 1, \cdots, n-1$. Let $a = (a_0, a_1, \ldots, a_{n-1}) \in C_1, b = (b_0, b_1, \ldots, b_{n-1}) \in C_2, c = (c_0, c_1, \ldots, c_{n-1}) \in C_3$. T is shift operator. If $C = \eta_1 C_1 \oplus \eta_2 C_2 \oplus \eta_3 C_3$ is a cyclic code over R, then $T(\alpha) = \eta_1 T(a) + \eta_2 T(b) + \eta_3 T(c) \in C$. Hence $T(a) \in C_1, T(b) \in C_2$ and $T(c) \in C_3$. On the other hand, if $C_i(i = 1, 2, 3)$ are cyclic codes over $\mathbb{F}_p$, then $T(a) \in C_1, T(b) \in C_2, T(c) \in C_3$. Hence $T(\alpha) = \eta_1 T(a) + \eta_2 T(b) + \eta_3 T(c) \in C$, which means that C is a cyclic code over R. Next, assuming that $C_i = (f_i')$, where $f_i' \in \mathbb{F}_p[x], f_i'|x^n - 1, |C_i| = p^{n - \deg(f_i')}, (i = 1, 2, 3)$. It is obvious $C \subseteq (\eta_1 f_1', \eta_2 f_2', \eta_3 f_3')$. Now, let $r = \eta_1 f_1' r_1 + \eta_2 f_2' r_2 + \eta_3 f_3' r_3 \subseteq (\eta_1 f_1', \eta_2 f_2', \eta_3 f_3')$, where $r, r_i \in R[x], i = 1, 2, 3$, so there exit $a_i', b_i', c_i' \in \mathbb{F}_p[x]$ such that $r_i = \eta_1 a_i' + \eta_2 b_i' + \eta_3 c_i'$. Hence $r = \eta_1 f_1' r_1 + \eta_2 f_2' r_2 + \eta_3 f_3' r_3 = \eta_1 f_1' a_1' + \eta_2 f_2' b_2' + \eta_3 f_3' c_3' \in C$. That is $(\eta_1 f_1', \eta_2 f_2', \eta_3 f_3') \subseteq C$.

Theorem 3.2. [8] Let C be a cyclic code of length n over R, then there exists a polynomial $g \in R[x]$ such that $C = (g)$ and $g|x^n - 1$.

Proof. By Theorem 3.1, we may assume that $C = (\eta_1 f_1', \eta_2 f_2', \eta_3 f_3')$, where $C_i = (f_i')(i = 1, 2, 3)$. Let $g = \eta_1 f_1' + \eta_2 f_2' + \eta_3 f_3'$. Clearly, $(g) \subseteq C$. On the other hand, $\eta_i f_i' = \eta_i g, i = 1, 2, 3$. This gives $C \subseteq (g)$, hence $C = (g)$.

According to Theorems 3.2, we have the following theorem, the proof of which follows by the Bezout identity like in the field alphabet case.

Theorem 3.3. Let $C = \eta_1 C_1 \oplus \eta_2 C_2 \oplus \eta_3 C_3$ be a cyclic code of length n over R, where $(n, p) = 1$, $C_i = (f_i)$, $f_i (i = 1, 2, 3)$ are idempotents, then there is only one idempotent $e \in C$ such that $C = (e)$, where $e = \eta_1 f_1 + \eta_2 f_2 + \eta_3 f_3$.

4   Quadratic Residue Codes over $\mathbb{F}_p + v\mathbb{F}_p + v^2\mathbb{F}_p$

In this section, quadratic residue codes over R are defined in terms of their idempotent generators. Let q be an odd prime such that $q \equiv \pm 1 \pmod 4$ and let Q_q and N_q be the sets of quadratic residues and non-residues modulo q, respectively. We use the notations $g_1 = \sum_{i \in Q_q} x^i$, $g_2 = \sum_{i \in N_q} x^i$ and h denotes the polynomial corresponding to the all one vector of length q, i.e. $h = 1 + g_1 + g_2$. Following classical notation [1] we let for $i = 1, 2$ the codes C_i (resp. C_i') be generated by g_i (resp. $(x + 1)g_i$).

Lemma 4.1. [1] If $p > 2$ and $q = 4k \pm 1$, then idempotent generators of quadratic residue codes C_1, C_1', C_2, C_2' over $\mathbb{F}_p$ are $E_q(x) = \frac{1}{2}(1 + \frac{1}{q}) + \frac{1}{2}(\frac{1}{q} - \frac{1}{\theta})g_1 + \frac{1}{2}(\frac{1}{q} + \frac{1}{\theta})g_2$, $F_q(x) = \frac{1}{2}(1 - \frac{1}{q}) - \frac{1}{2}(\frac{1}{q} + \frac{1}{\theta})g_1 - \frac{1}{2}(\frac{1}{q} - \frac{1}{\theta})g_2$, $E_n(x) = \frac{1}{2}(1 + \frac{1}{q}) + \frac{1}{2}(\frac{1}{q} - \frac{1}{\theta})g_2 + \frac{1}{2}(\frac{1}{q} + \frac{1}{\theta})g_1$, $F_n(x) = \frac{1}{2}(1 - \frac{1}{q}) - \frac{1}{2}(\frac{1}{q} + \frac{1}{\theta})g_2 - \frac{1}{2}(\frac{1}{q} - \frac{1}{\theta})g_1$, respectively, where θ denotes *Gaussian* sum, $\chi(i)$ denotes *Legendre* symbol, that is,

$$\theta = \sum_{i=1}^{q-1} \chi(i)\alpha^i, \quad \chi(i) = \begin{cases} 0, & p \mid i; \\ 1, & i \in Q_q; \\ -1, & i \in N_q, \end{cases}$$

where α is a primitive q^{th} root of unit over some extension field of $\mathbb{F}_p$. For convenience, we set $e_1 = E_q(x)$, $e_1' = F_q(x)$, $e_2 = E_n(x)$, $e_2' = F_n(x)$.

Using Lemmas 2.1 and 4.1, we can obtain the following theorem, which plays an important role in the main results.

Lemma 4.2. $\eta_1 e_i + \eta_2 e_j + \eta_3 e_k$, $\eta_1 e_i' + \eta_2 e_j' + \eta_3 e_k'$ are idempotents in R_q, where e_i, e_j, e_k are not all equal, e_i', e_j', e_k' are not all equal $(i, j, k = 1, 2)$.

Next, define the quadratic residue codes over R in terms of their idempotent generators.

Definition 4.3. Let q be an odd prime such that p is a quadratic residue modulo q. Set
$Q_3 = (\eta_1 e_2 + \eta_2 e_1 + \eta_3 e_1)$, $Q_2 = (\eta_1 e_1 + \eta_2 e_2 + \eta_3 e_1)$, $Q_3 = (\eta_1 e_1 + \eta_2 e_1 + \eta_3 e_2)$,
$Q_4 = (\eta_1 e_1 + \eta_2 e_2 + \eta_3 e_2)$, $Q_5 = (\eta_1 e_2 + \eta_2 e_1 + \eta_3 e_2)$, $Q_6 = (\eta_1 e_2 + \eta_2 e_2 + \eta_3 e_1)$,
$S_1 = (\eta_1 e_2' + \eta_2 e_1' + \eta_3 e_1')$, $S_2 = (\eta_1 e_1' + \eta_2 e_2' + \eta_3 e_1')$, $S_3 = (\eta_1 e_1' + \eta_2 e_1' + \eta_3 e_2')$,
$S_4 = (\eta_1 e_1' + \eta_2 e_2' + \eta_3 e_2')$, $S_5 = (\eta_1 e_2' + \eta_2 e_1' + \eta_3 e_2')$, $S_6 = (\eta_1 e_2' + \eta_2 e_2' + \eta_3 e_1')$.
These twelve codes are called quadratic residue codes over R of length q.

Note that $\sum_{i=1}^{3}\eta_i = 1$, we have $Q_i = ((1-\eta_i)e_1 + \eta_i e_2)$, $Q_{3+i} = ((1-\eta_i)e_2 + \eta_i e_1)$, $S_i = ((1-\eta_i)e_1' + \eta_i e_2')$, $S_{3+i} = ((1-\eta_i)e_2' + \eta_i e_1')(i = 1, 2, 3)$.

As in the case of quadratic residue codes over finite ring $\mathbb{F}_2 + u\mathbb{F}_2 + u^2\mathbb{F}_2$, the properties of quadratic residue codes over R differ from the cases $q \equiv 3 \pmod 4$ and $q \equiv 1 \pmod 4$.

Theorem 4.4. If $q \equiv 3 \pmod 4$, with the notation as in Definition 4.3, the following assertions hold for quadratic residue codes over R:

(a) Q_i and S_i are equivalent to Q_{3+i} and S_{3+i}, respectively, $i = 1, 2, 3$;

(b) $Q_i \cap Q_{3+i} = (\frac{1}{q}h)$, $Q_i + Q_{3+i} = R_q, i = 1, 2, 3$;

(c) $Q_j = S_j \mid (\frac{1}{q}h), j = 1, 2, 3, 4, 5, 6$;

(d) $|Q_j| = (p^3)^{(q+1)/2}, |S_j| = (p^3)^{(q-1)/2}, j = 1, 2, 3, 4, 5, 6$;

(e) S_j are self-orthogonal and $Q_j^{\perp} = S_j, j = 1, 2, 3, 4, 5, 6$;

(f) $S_i \cap S_{3+i} = \{0\}, S_i + S_{3+i} = (1 - \frac{1}{q}h), i = 1, 2, 3$.

Proof. (a) Let $n \in N_q$, then $\mu_n(g_1) = g_2$ and $\mu_n(g_2) = g_1$. Hence $\mu_n(e_1) = e_2$, $\mu_n(e_2) = e_1$, $\mu_n(e_1') = e_2'$, $\mu_n(e_2') = e_1'$. Therefore, $\mu_n((1-\eta_i)e_1 + \eta_i e_2) = (1-\eta_i)e_2 + \eta_i e_1, \mu_n((1-\eta_i)e_1' + \eta_i e_2') = (1-\eta_i)e_2' + \eta_i e_1'$, which implies Q_i and Q_{3+i} are equivalent, S_i and S_{3+i} are equivalent.

(b) By direct calculation, we have $((1-\eta_i)e_1 + \eta_i e_2) \cdot ((1-\eta_i)e_2 + \eta_i e_1) = ((1-\eta_i)e_1 + \eta_i e_2) \cdot [((1-\eta_i)e_1 + \eta_i e_2) + ((1-\eta_i)e_2 + \eta_i e_1) - 1] = ((1-\eta_i)e_1 + \eta_i e_2) \cdot (e_1 + e_2 - 1) = ((1-\eta_i)e_1 + \eta_i e_2) \cdot \frac{1}{q}h = \frac{1}{q} \cdot \{(1-\eta_i)[\frac{1}{2}(1 + \frac{1}{q})h + \frac{1}{2}(\frac{1}{q} - \frac{1}{\theta})\frac{q-1}{2}h + \frac{1}{2}(\frac{1}{q} + \frac{1}{\theta})\frac{q-1}{2}h] + \eta_i[\frac{1}{2}(1 + \frac{1}{q})h + \frac{1}{2}(\frac{1}{q} + \frac{1}{\theta})\frac{q-1}{2}h + \frac{1}{2}(\frac{1}{q} - \frac{1}{\theta})\frac{q-1}{2}h]\} = \frac{1}{q}h$. According to Theorem 2.2, we have $Q_i \cap Q_{3+i} = (\frac{1}{q}h)$. $((1-\eta_i)e_1 + \eta_i e_2) + ((1-\eta_i)e_2 + \eta_i e_1) + ((1-\eta_i)e_1 + \eta_i e_2) \cdot ((1-\eta_i)e_2 + \eta_i e_1) = e_1 + e_2 - \frac{1}{q}h = 1$. Hence $Q_i + Q_{3+i} = R_q$.

(c) By direct calculation, we have $((1-\eta_i)e_1' + \eta_i e_2') \cdot \frac{1}{q}h = \frac{1}{q} \cdot \{(1-\eta_i)[\frac{1}{2}(1 - \frac{1}{q})h - \frac{1}{2}(\frac{1}{q} + \frac{1}{\theta})\frac{q-1}{2}h - \frac{1}{2}(\frac{1}{q} - \frac{1}{\theta})\frac{q-1}{2}h] + \eta_i[\frac{1}{2}(1 - \frac{1}{q})h - \frac{1}{2}(\frac{1}{q} - \frac{1}{\theta})\frac{q-1}{2}h - \frac{1}{2}(\frac{1}{q} + \frac{1}{\theta})\frac{q-1}{2}h]\} = 0 = ((1-\eta_i)e_2' + \eta_i e_1') \cdot \frac{1}{q}h$. Therefore, $S_j \cap (\frac{1}{q}h) = 0$. Since $((1-\eta_i)e_1' + \eta_i e_2') + \frac{1}{q}h = ((1-\eta_i)e_1' + \eta_i e_2') + ((1-\eta_i) + \eta_i)\frac{1}{q}h = ((1-\eta_i)e_1 + \eta_i e_2) \cdot ((1-\eta_i)e_2' + \eta_i e_1') + \frac{1}{q}h = ((1-\eta_i)e_2 + \eta_i e_1)$ Hence, $Q_j = S_j + (\frac{1}{q}h)$.

(d) According to (a) and (b), we have $|Q_i \cap Q_{3+i}| = |(\frac{1}{q}h)| = p^3$. Since $(p^3)^q = |Q_i + Q_{3+i}| = \frac{|Q_i||Q_{3+i}|}{|Q_i \cap Q_{3+i}|} = \frac{|Q_i|^2}{p^3}$. Thus, $|Q_i| = |Q_{3+i}| = (p^3)^{(q+1)/2}$. Since $(p^3)^{(q+1)/2} = |Q_j| = |S_j + (\frac{1}{q}h)| = |S_j||(\frac{1}{q}h)| = p^3|S_j|$, we have $|S_j| = (p^3)^{(q-1)/2}$.

(e) Since $-1 \in N_q$, according to Theorem 2.3, the generating idempotent of $C_1^{\perp}$ of $C_1 = (e_1)$ is $1 - e_1(x^{-1}) = 1 - \frac{1}{2}(1 + \frac{1}{q}) - \frac{1}{2}(\frac{1}{q} - \frac{1}{\theta})g_2 - \frac{1}{2}(\frac{1}{q} + \frac{1}{\theta})g_1 = \frac{1}{2}(1 - \frac{1}{q}) - \frac{1}{2}(\frac{1}{q} + \frac{1}{\theta})g_1 - \frac{1}{2}(\frac{1}{q} - \frac{1}{\theta})g_2 = e_1'$. Similarly, the generating idempotent of $C_2^{\perp}$ is e_2'. Using Theorem 2.3, the generating idempotent of $Q_i^{\perp}(i = 1, 2, 3)$ are $(1 - \eta_i)e_1' + \eta_i e_2'$, which implies $Q_i^{\perp} = S_i$. According to c), we have $S_i \subseteq Q_i = S_i^{\perp}$. Hence, $S_i(i = 1, 2, 3)$ are self-orthogonal. Similarly, $Q_{3+i}^{\perp} = S_{3+i}$ and $S_{3+i}(i = 1, 2, 3)$ are self-orthogonal.

(f) Since $((1-\eta_i)e_1' + \eta_i e_2') \cdot ((1-\eta_i)e_2' + \eta_i e_1') = ((1-\eta_i)e_1' + \eta_i e_2') \cdot [((1-\eta_i)e_1' + \eta_i e_2') + ((1-\eta_i)e_2' + \eta_i e_1') - 1] = ((1-\eta_i)e_1' + \eta_i e_2') \cdot (e_1' + e_2' - 1) =$

$((1 - \eta_i)e_1' + \eta_i e_2') \cdot (-\frac{1}{q}h) = 0$. That is, $S_i \cap S_{3+i} = 0$. $((1 - \eta_i)e_1' + \eta_i e_2') + ((1 - \eta_i)e_2' + \eta_i e_1') = e_1' + e_2' = 1 - \frac{1}{q}h$. Hence, $S_i + S_{3+i} = (1 - \frac{1}{q}h)$.

Similar to the proof of Theorem 4.4, we have the following theorem.

Theorem 4.5. If $q \equiv 1 \pmod 4$, with the notation as in Definition 4.3, then the following assertions hold for quadratic residue codes over R:

(a) Q_i and S_i are equivalent to Q_{3+i} and S_{3+i}, respectively, $i = 1, 2, 3$;
(b) $Q_i \cap Q_{3+i} = (\frac{1}{q}h)$, $Q_i + Q_{3+i} = R_q, i = 1, 2, 3$;
(c) $Q_j = S_j + (\frac{1}{q}h), j = 1, 2, 3, 4, 5, 6$;
(d) $|Q_j| = (p^3)^{(q+1)/2}, |S_j| = (p^3)^{(q-1)/2}, j = 1, 2, 3, 4, 5, 6$;
(e) $Q_i^{\perp} = S_{3+i}, Q_{3+i}^{\perp} = S_i, i = 1, 2, 3$;
(f) $S_i \cap S_{3+i} = 0, S_i + S_{3+i} = (1 - \frac{1}{q}h), i = 1, 2, 3$.

5   Extended Quadratic Residue Codes over $\mathbb{F}_p + v\mathbb{F}_p + v^2\mathbb{F}_p$

In this section, we discuss the properties of extended quadratic residue codes over R.

Definition 5.1. The extended code of a code C over R will be denoted by $\overline{C}$, which is the code obtained by adding a specific column to the generator matrix of C. In addition, define the generator matrix of $\widehat{Q}_j$ as

$$
\begin{array}{ccccccc}
\infty & 0 & 1 & 2 & \cdots & q-1 &
\end{array}
$$
$$
\left(
\begin{array}{cccccc}
0 & & & & & \\
0 & & & G_j' & & \\
\vdots & & & & & \\
1 & 1 & 1 & 1 & \cdots & 1
\end{array}
\right),
$$

where G_j' generates $S_j (j = 1, 2, \cdots 6)$, and the row above the horizontal bar shows the column labelling by $\mathbb{F}_q \cup \infty$.

Theorem 5.2. If $q \equiv 3 \pmod 4$, with the notation $Q_j (j = 1, 2, \cdots 6)$ as in Definition 4.3, then $\overline{Q}_j^{\perp} = \widehat{Q}_j$. In particular, if $1 + \frac{1}{q} = 0$, then $\overline{Q}_j$ are self-dual.

Proof. Theorem 4.4 tells us that $Q_j = S_j + (\frac{1}{q}h)(j = 1, 2, 3, 4, 5, 6)$, then the generator matrix of $\overline{Q}_j$ are

$$
\begin{array}{ccccccc}
\infty & 0 & 1 & 2 & \cdots & q-1 &
\end{array}
$$
$$
\left(
\begin{array}{cccccc}
0 & & & & & \\
0 & & & G_j' & & \\
\vdots & & & & & \\
-1 & \frac{1}{q} & \frac{1}{q} & \frac{1}{q} & \cdots & \frac{1}{q}
\end{array}
\right),
$$

where G'_j are a generator matrix of S_j. Since S_j are self-orthogonal, any two rows of G'_j are orthogonal. According to the proof of (c) in Theorem 4.4, we know that each line of G'_j are orthogonal together with the vector $(\frac{1}{q}h)$. Since $(1,h) \cdot (-1, \frac{1}{q}h) = 0$, then $|\overline{Q}_j^\perp| = |\widehat{Q}_j| = (p^3)^{(q+1)/2}$. That is, $\overline{Q}_j^\perp = \widehat{Q}_j$. In particular, if $1 + \frac{1}{q} = 0$, $\overline{Q}_j$ are linear codes generated by the matrix

$$
\overline{G}_j = \begin{matrix} & \infty & 0 & 1 & 2 & \cdots & q-1 \\ & \begin{pmatrix} 0 & & & & & \\ 0 & & & & G'_j & \\ \vdots & & & & & \\ -1 & -1 & -1 & -1 & \cdots & -1 \end{pmatrix} \end{matrix},
$$

Obviously, $(1,h) \in \overline{G}_j$. Hence, $\overline{Q}_j^\perp = \overline{Q}_j$. That is, $\overline{Q}_j$ are self-dual.

Similar to the proof of Theorem 5.2, we have the following theorem.

Theorem 5.3. If $q \equiv 1 \pmod 4$, with the notation $Q_j(j = 1, 2, \cdots 6)$ as in Definition 4.3, then $\overline{Q}_i^\perp = \widehat{Q}_{3+i}, \overline{Q}_{3+i}^\perp = \widehat{Q}_i (i = 1, 2, 3)$. In particular, if $1 + \frac{1}{q} = 0$, then $\overline{Q}_i^\perp = \overline{Q}_{3+i}, \overline{Q}_{3+i}^\perp = \overline{Q}_i (i = 1, 2, 3)$.

6   Numerical Examples

In this section, we give some examples to validate the main conclusions obtained in this paper. The parameters given are that of the finite field image defined for all $a_i \in \mathbb{F}_p$ by the formula

$$
\Phi\left(\sum_{i=1}^{3} a_i \eta_i\right) = (a_1 + a_3, a_1 + a_2, a_2 + a_3).
$$

Note that ϕ does not map self-dual codes to self dual codes, but it does give some good Hamming distance codes.

- Let $p = 3$ and $q = 11$. The sets of quadratic residues and non-residues modulo q are $Q_q = \{1, 3, 4, 5, 9\}$, and $N_q = \{2, 6, 7, 8, 10\}$, respectively. Thus, $g_1(x) = \sum_{i \in Q_q} x^i = x + x^3 + x^4 + x^5 + x^9$ and $g_2(x) = \sum_{i \in N_q} x^i = x^2 + x^6 + x^7 + x^8 + x^{10}$. According to Lemma 4.1, $e'_1 = 1 + g_2$, $e'_2 = 1 + g_1$, and $e_i = -g_i$, where $i = 1, 2$. By Definition 4.3, $Q_3 = (-v^2(x + x^3 + x^4 + x^5 + x^9) - (1 - v^2)(x^2 + x^6 + x^7 + x^8 + x^{10}))$ and $S_3 = (v^2(1 + x^2 + x^6 + x^7 + x^8 + x^{10}) + (1 - v^2)(1 + x + x^3 + x^4 + x^5 + x^9))$. Theorem 4.4 tells us that S_3 is self-orthogonal and $Q_3^\perp = S_3$. According to Theorem 5.2, $\overline{Q}_3$ is self-dual. It image by ϕ has parameters $[36, 18, 11]$.
- Let $q = 3$ and $p = 7$ or $p = 13$. The code $\phi(Q_1)$ is an almost MDS $[9, 6, 3]$ code in the sense of [9]. It is not nearly MDS since the dual has parameters $[9, 3, 3]$.
- Let $q = 23$ and $p = 3$. The code $\phi(\overline{Q}_1)$ has parameters $[72, 36, 13]$.

7 Conclusion

This article gives the definition and some properties of quadratic residue codes over the ring $\mathbb{F}_p + v\mathbb{F}_p + v^2\mathbb{F}_p$, subject to the restriction $v^3 = v$, where p is an odd prime. In [7], the quadratic residue codes over $\mathbb{F}_2 + v\mathbb{F}_2 + v^2\mathbb{F}_2 (v^3 = v)$ are studied in detail. In this paper, we generalize the structural results of [7] by replacing $\mathbb{F}_2$ by $\mathbb{F}_p$, for odd primes p. We derived the generating idempotents of quadratic residue codes. Some Gray images turn out to be almost MDS. Alternative Gray maps compatible with duality and/or cyclicity are worth exploring.

References

1. MacWilliams, F.J., Sloane, N.J.A.: The Theory of Error-Correcting Codes. North Holland Publishing Co, Amsterdam (1977)
2. Pless, V., Qian, Z.: Cyclic codes and quadratic residue codes over $\mathbb{Z}_4$. IEEE Trans. Inform. Theory **42**(5), 1594–1600 (1996)
3. Tan, X.: A family of quadratic residue codes over $\mathbb{Z}_{2^m}$. In: Fourth International Conference on Emerging Intelligent Data Web Technologies, pp. 236–240 (2013)
4. Kaya, A., Yildiz, B., Siap, I.: Quadratic Residue Codes over $\mathbb{F}_p + v\mathbb{F}_p$ and their Gray Images. arXiv preprint arXiv: 1305, 4508 (2013)
5. Zhang, T., Zhu, S.X.: Quadratic residue codes over $\mathbb{F}_p + v\mathbb{F}_p$. J. Univ. Sci. Technol. China **42**(3), 208–213 (2012). In Chinese
6. Shi, M.J., Solé, P., Wu, B.: Cyclic codes and the weight enumerator of linear codes over $\mathbb{F}_2 + v\mathbb{F}_2 + v^2\mathbb{F}_2$. Appl. Comput. Math **12**(2), 247–255 (2013)
7. Kaya, A., Yildiz, B., Siap, I.: New extremal binary self-dual codes of length 68 from quadratic residue codes over $\mathbb{F}_2 + u\mathbb{F}_2 + u^2\mathbb{F}_2$. Finite Fields Appl. **29**, 160–177 (2014)
8. Cao, D.C., Zhu, S.X.: Constacyclic codes over the ring $\mathbb{F}_q + v\mathbb{F}_q + v^2\mathbb{F}_q$. J. Hefei Univ. Technol. **36**(12), 1534–1536 (2013). In Chinese
9. de Boer, M.: Almost MDS codes. Des. Codes Crypt. **9**(2), 143–155 (1996)

Author Index

Printed in the United States
By Bookmasters